COSMOS, CHAOSMOS AND ASTROLOGY

Cosmos, Chaosmos and Astrology:
Rethinking the Nature of Astrology

SOPHIA CENTRE MASTER MONOGRAPHS: VOLUME 1

Sophia Centre for the Study of Cosmology in Culture
University of Wales Trinity St David
Jennifer Zahrt, General Editor

Cosmos, Chaosmos and Astrology
RETHINKING THE NATURE OF ASTROLOGY

by Bernadette Brady

SOPHIA CENTRE PRESS

First published by Sophia Centre Press in 2014.

SOPHIA CENTRE PRESS
University of Wales Trinity Saint David
Ceredigion, Wales SA48 7ED, United Kingdom
www.sophiacentrepress.com

ISBN 978-1-907767-40-1
British Library Cataloguing in Publication Data.
A catalogue card for this book is available from the British Library.

Book Design by Joseph Uccello.

Printed in the UK by LightningSource.

ABSTRACT

This work explores the possible links between the practice of contemporary astrology with findings of chaos theory and complexity science. In this regard it joins other academic endeavours which have explored the chaosmological nature of different forms of literature, music, philosophy, and religion. Extensive as these other works are, as yet no research has been done to view the enigmatic subject of astrology in this light. The work considers a number of parallel paths. These paths are: creation mythology or vernacular ontologies; the philosophical framework of the period of the origins of astrology; and the implications of the relevance of chaos theory and complexity science in the human sciences. With these threads in place I map some of the philosophy and practices of astrology onto the findings of chaos theory and complexity. The work concludes by offering a view of astrology which is neither a pseudo-science looking for a causal agent nor a subject that requires a spiritual component. Additionally, while acknowledging a level of cultural relativism to astrology, I argue that there is also a level of philosophical absolutism evident in the subject's consistent approach to its views on the union of sky and life. The work concludes by viewing astrology as a product of human intuition put to the service of the humanity's need to bring a level of domestication to chaos in order to give meaning to life. In this regard it can be considered as one of humanity's enduring subjects.

For
IRENE EARIS
(1946–2013)

ACKNOWLEDGEMENTS

I would like to thank Prof. Michael York who first introduced me to the thinking of Complexity and Patrick Curry for his encouragement and introducing me to the world of chaotic myths. Additionally, I would like to acknowledge the influence on my thinking of Nicholas Campion with whom I have had many hours of conversations around the nature of astrology. I would also like to thank Darrelyn Gunzburg for her readings, comments to various versions of this work, and her support, which has enabled my thinking to find its voice.

TABLE OF CONTENTS

LIST OF FIGURES

CHAPTER 1:

INTRODUCTION

THE POTENTIAL LINKS between the practice of contemporary astrology and the findings of chaos theory and
complexity science are the focus of this work. In this regard it joins other academic attempts which have explored
the chaosmological nature of different forms of literature,
music, philosophy, and religion.[1] Extensive as these other

1 See for example, Cristina Farronato, *Eco's Chaosmos: From the Middle Ages
to Postmodernity* (Toronto; London: University of Toronto Press, 2003); Tim
Clark, 'A Whiteheadian Chaosmos: Process Philosophy from a Deleuzean
Perspective', *Process Studies* 3–4 (1999): pp. 179–94; Frederick J. Ruf, *The
Creation of Chaos: William James and the Stylistic Making of a Disorderly
World* (Albany: State University of New York Press, 1991); Jean-Godefroy
Bidima, 'Music and the Socio-historical Real', in *Deleuze and Music*, ed. by
Ian Buchanan and Marcel Swiboda (Edinburgh: Edinburgh University Press,
2004), pp. 176–95.

works are, as yet, no research has been undertaken with a view to considering the subject of astrology in this light.

The trajectory of this work is to consider a number of parallel paths: creation mythology or vernacular ontologies; the philosophical orientation of the culture that placed astrology in its national thinking; and the implications of the relevance of chaos theory and complexity science in the human sciences. With these threads in place I will follow the leads of Cristiana Farronato when considering Umberto Eco's literary work, Tim Clark when considering Alfred Whitehead's philosophy, Fredrick Ruf when considering the philosophy of William James, and Jean-Godefroy Bidima when considering Gilles Deleuze's comments on music. Applying a similar line of thinking to astrology I argue that the phenomenon of a lived life has given rise to a vernacular set of astrological tenets and practices which mirror, even if somewhat crudely, the current findings within chaos theory and complexity science. Accordingly my research offers a view on the nature of astrology suggesting that it is neither a pseudo-science nor a spiritual subject but is instead a product of the human desire to domesticate chaos, a blending of the chaosmic notion of *sumpatheia* with the cosmic surety of heavenly numbers.

This first chapter establishes the attributes of the three central themes of the work: chaosmos, and the notion of *sumpatheia*; cosmos; and astrology.

THE NATURE OF CHAOSMOS

It was Hesiod (c. 750-650 BCE) who offered the first known comments on Chaos (the place). He spoke of Chaos in his creation myth as, 'verily at the first Chaos came to be', thus naming it as the first thing or place that existed and from Chaos the gods emerged, first Erebus and the Black Night and then the Day.[2] For Hesiod the gods, once born, stayed away from 'gloomy Chaos' living instead in the glorious heavens propped up with 'silver pillars'.[3] Later Plato (c. 427-348 BCE) carried this idea forward and defined his cosmos as the outcome of the bringing of order to disorder, 'because he [the god] believed that order was in every way better than disorder'.[4] In this regard the Greek concepts of *Kosmos* and Chaos were a pair involved in a creative union. *Kosmos* came from Chaos and was beautiful order, the order which is understandable, knowable, reliable and, therefore, defined by Plato as containing Reason. In contrast, *Kosmos'* non-gendered parent, Chaos, brought to mind infinite and unending space, the abyss or gulf with no limits, the place from which elements of the *Kosmos* fought to be released, or to crawl free, either by natural emergence, or as was the case in the Babylonian creation mythology of Marduk and

---

2 Hesiod, *Works and Days, Theogony and the Shield of Hercules* (New York: Dover Publications Inc., 2006), p. 32.

3 Hesiod, *Works and Days*, p. 49.

4 Plato, *Timaeus*, trans. by Donald J. Zeyl, in *Plato: Complete Works*, ed. by John M. Cooper, (Cambridge: Hackett Publishing Company, 1997), pp. 1224–91, here, line 30a.

the chaotic force of Tiamat, by force and bloodshed.[5] In this context Chaos became defined as disorder, and disorder remained the dominant definition of Chaos from the classical period to the twentieth century.

Both *Kosmos* and Chaos were also 'located' in the heavens; *Kosmos* touched the earth through heavenly order and Chaos by heavenly disorder. The continuity of order, however, was not assumed and the sky needed to be watched for any signs of the emergence of Chaos. Many letters written to the Assyrian kings in the seventh century BCE by their astronomer/astrologer/priests were concerned with the phases of the moon and its regularity against their established order of the lunar calendar. An example of such letters stated that, 'if the moon becomes visible on the 1st day: reliable speech; the land will become happy' or in contrast, 'if the moon becomes visible on the 30th day there will be a rumour of the enemy'.[6] Every month a priest needed to observe the full moon for, 'if the moon and the sun are in balance: the land will become stable...if on the 14th day the moon and the sun are seen together [in the sky]: reliable speech, the land will become happy'.[7] In contrast, the priest Issar-sumu-eres warned the king that, 'If the moon does not wait for the sun but sets: raging of lion and wolf'.[8] Ritu-

5 L. W. King, *The Seven Tablets of Creation: or, The Babylonian and Assyrian Legends Concerning the Creation of the World and of Mankind* (1902; repr. Montana: Kessinger Publishing, 2004), Tablet IV, line 56–60.

6 Hermann Hunger, *Astrological Reports to Assyrian Kings* (Helsinki: Helsinki University Press, 1992), p. 10, trans. 10 & 11.

7 Hunger, *Astrological Reports to Assyrian Kings*, p. 11, trans. 15.

8 Hunger, *Astrological Reports to Assyrian Kings*, p. 15, trans. 24.

als could be used to minimise or neutralise the rumblings of chaos, and thus the astronomer/astrologer/priests were sentries standing on the edge of, and guarding, the ordered world.

Nearly two thousand years later when the heavenly order was accepted as the norm, the idea of chaos (the state) was still associated with a disordered heaven. The Welsh historical commentator, Geoffrey of Monmouth (c. 1100–1155), described the zodiac signs losing their order as the herald of approaching chaos,

> The Twins shall surcease from their wonted embrace, and shall call the Urn unto the fountains. The scales of the Balance shall hang awry until the Ram shall set his crooked horns beneath them. The tail of the Scorpion shall breed lightnings, and the Crab fall at strife with the Sun. The Virgin shall forget her maiden shame, and climb up on the back of the Sagittary. The chariot of the Moon shall disturb the Zodiac, and the Pleiades shall burst into tears and lamentation.[9]

The sky in disorder was also William Shakespeare's (1564–1616) omen for the eruption of chaos. In his play *Troilus and Cressida* he described such a time as,

> Sans check, to good and bad: but when the planets
> In evil mixture to disorder wander,

9 Geoffrey of Monmouth, *History of the Kings of Britain* (London: J. M. Dent and Co, 1904), p. 188.

What plagues, and what portents, what mutiny,
What raging of the sea, shaking of earth.[10]

This tradition of chaos as disorder infers that chaos is a state that once was ordered or may potentially be ordered in the future, thereby it implies that chaos and cosmos are two ends of a single continuum, one end being the dark void of disorder and the other, the bright clear world of order. Such a definition, however, denies the possibility of chaos holding any unique features.

It was the author James Joyce (1881–1941) who challenged this definition of chaos by suggesting that it was its own unique ontological force. In *Finnegans Wake* he introduced the term chaosmos, using it to describe the internal world of one of his characters named Alle, '...every person, place and thing in the chaosmos of Alle anyway connected with the gobblydumped turkery was moving and changing every part of the time'.[11] Philip Kuberski argued that Joyce used the term chaosmos to define a state of creation which was a self-organising order, order that emerges spontaneously from a void or chaotic state.[12] Kuberski added that the Joycean chaosmos 'could be described as a lexical and syntactic interiorzation of the flux of material existence' and that it was 'the world of objects moving in space into a constant flux and field of ever-new reorientations and re-

10 Shakespeare, *Troilus and Cressida*, Act I, Scene 3.

11 James Joyce, *Finnegans Wake* (1939; repr. Oxford: Oxford University Press, 2012), p. 118, lines 18–23.

12 Philip Kuberski, *Chaosmos: Literature, Science, and Theory* (Albany: State University of New York Press, 1994), pp. 37–38.

lationships'.[13] Kuberski observed how, by writing his prose in such an interconnected, chaosmotic manner, Joyce's text took on an extra dimension 'moving towards a nonlinear, nonidealized, ragged enactment of holism, the chaosmic text does not so much break with realism as deepen it, make it more capacious, able to play up and down various scales of size and perspective simultaneously, make it less easily satisfied with closure'.[14] Chaosmos for Joyce was thus the instrument used to give his prose complexity, a holism which gained depth by reflecting across scales.

Joyce's 'field of ever-new reorientations and relation-ships' is similar to the classical notion of Stoic *sumpatheia* or as the Pythagoreans named it, *harmonia*. Suzanne Bobzien wrote of this third century BCE Stoic idea of *sumpatheia* that it was the simple idea that all things influence each other.[15] Diogenes Laërtius in the third century CE described such influences as *all* things being 'rendered continuous by their mutual interchange'.[16] The universality of the notion that all things influence each other should not be ignored in these statements, as this universality promises a fully in-terlinked world which justifies the logic of omens or signs. A. A. Long and D. N. Sedley stated that a 'sign, according to Stoic doctrine, is an evident truth by which some further,

13 Kuberski, *Chaosmos: Literature, Science, and Theory*, p. 84.
14 Kuberski, *Chaosmos: Literature, Science, and Theory*, p. 192.
15 Susanne Bobzien, *Determinism and Freedom in Stoic Philosophy* (Oxford: Clarendon Press, 1998), p. 169.
16 Diogenes Laëtius, 'Physics', in *The Stoics Reader*, ed. by Brad Inwood and Lloyd P. Gerson (Cambridge, MA: Hackett Publishing Company, Inc., 2008), pp. 51–112, paragraph 84.

non-evident truth is revealed'.[17] Joyce, two thousand years later, used this connectedness, fuelled by non-evident associations, to deepen his prose. As will be discussed later, this connectedness is also a feature of the creative properties of the void of Chaos.

By way of contrast, Frederick Ruf suggested that for William James (1842–1910), chaosmos was the absence of cosmos. While James was returning to a more classical definition of an absence of order, Ruf asserted that this may only be an attempt by James to give form to a formless concept or to make chaosmos more mentally conceivable.[18] Umberto Eco's philosophy also appears to express this form of chaosmos. Cristina Farronato argued Eco's philosophy was similar to James', in that for Eco the 'tension between order and disorder, between cosmos and chaos, [brought] him to find a middle theory that characterizes his chaosmos'.[19] Eco's form of chaosmos was revealed in the nature of the library in his novel *The Name of the Rose*. The library was held together by non-sensible shifting relationships between books but the human need to find order sought to give it shape and pattern. Such desire Farronato described as a 'wishful longing, an elegant hope to see a method in the madness, or at least in the eternal return of the same

---

17 A. A. Long and D. N. Sedley, *The Hellenistic Philosophers*, Vol. 1 (1987; repr. Cambridge: Cambridge University Press, 2007), p. 264.

18 Ruf, *The Creation of Chaos*, pp. 9–10.

19 Farronato, *Eco's Chaosmos*, p. 11.

disorder'.[20] Thus for Eco chaosmos is real and cosmic order is an illusion.

The French philosopher Gilles Deleuze (1925–1995) stressed that chaosmos was cyclic and he defined it as follows, 'The eternal return is not the effect of the Identical upon a world become similar, it is not an external order imposed upon the chaos of the world; on the contrary, the eternal return is the internal identity of the world and of chaos, the Chaosmos'.[21] Jean-Godefroy Bidima saw Deleuze's chaosmos as the same as that of Joyce which Bidmia viewed as 'counter all Hellenic and pacific views of the "cosmos" as being the well-ordered, immutable and sufficient One against the Multiple'.[22] I would add, however, that such a chaosmos is a Multiple without definition or distinction. It is along these lines that the ideas of process philosophy of 'becoming' rather than 'being', as articulated by Alfred Whitehead (1861–1947), provide another view of chaosmos. Whitehead's chaosmos was summarised by Tim Clark as 'neither stable nor unstable, but "metastable," endowed with a potential energy'.[23]

Taking all of these definitions into account chaosmos can be considered as: a multiple without definition or dis-

---

20 Farronato, 'From the Rose to the Flame: Eco's theory and fiction between the Middle Ages and postmodernity', in *New Essays on Umberto Eco*, ed. by Peter E. Bondanella (Cambridge: Cambridge University Press, 2009), pp. 50–70, here p. 59.

21 Gilles Deleuze, *Difference and Repetition* (London: Continuum, 2004), p. 372.

22 Bidima, 'Music and the Socio-historical Real', p. 192.

23 Clark, 'A Whiteheadian Chaosmos', p. 193.

tinction, cyclic but not the return of the same, indifferent to scale or form, and metastable, a type of stability which is a potential buried within the ontological force of the void. This potential of the void, its creative force, is produced by it being impregnated with a web of connectivity which binds *all* to *everything*, (*sumpatheia*). These are the attributes of chaosmos that I use in this work which provide the premise for insights into the experience of living. For, following the lead of James Joyce, a life lived in chaosmos becomes the collision of the many. Such collisions produce the experience of life as a kaleidoscope of cyclic potentials, ideas, coincidences, and events which can fracture across a plethora of reflective 'surfaces' as signs, omens, and coincidences, all of which deepen and enrich the individual's life.

THE NATURE OF COSMOS

In contrast to chaosmos the classical cosmos (*Kosmos*) was a world born from a stable state of intellect and logical order. Plato, in *Gorgias*, defined *Kosmos* as the 'partnership and friendship, orderliness, self-control, and justice' which 'hold together heaven and earth, and gods and men, and that is why they call this universe *a world order* (*kosmos*), my friend, and not an undisciplined world-disorder'.[24] Edward Casey noted that for Plato in the *Timaeus* 'the special power of mathematics to shape a cosmos proceeds from the

---

24 Plato, *Gorgias*, 508a.

sky downward'.[25] In this way Plato's cosmos was also held together by a geometrically constructed World Soul which was driven by a Divine Intellect, which then shaped the world.[26] It is tempting to associate Plato's World Soul with the later Stoic notion of *sumpatheia* but, as Frances Cornford noted, Plato's World Soul was a hybrid of both perfect 'being' and imperfect 'becoming', and as such all human souls in the imperfect 'becoming' could partake in the World Soul located in the perfect 'being'.[27] Thus this union was not of equals, for Plato's notion of interconnectedness was hierarchical, with the privileging of God and Reason.[28] Indeed Plotinus (204/5-270), following Plato's thinking, called it the *kosmos noêtos*, the intelligent cosmos, with the assumption that this intelligence was driven and dictated by the One.[29] Plato's cosmology therefore is hierarchical, orderly, logical, designed using numbers by an artisan god, and maintained by a sky-based divine mind.

By the twentieth century Plato's sky-based view of cosmology had morphed into large-scale astronomy with divine Reason being replaced with the order of the laws of physics. Recently, however, cosmology has come down to earth. The notion of cosmos or Kosmos, has now shifted in

25 Edward S. Casey, *The Fate of Place: A Philosophical History* (Berkeley, CA: University of California Press, 1998), p.39.

26 Plato, *Timaeus*, 34a-b.

27 Francis M. Cornford, *Plato's Cosmology* (1935; repr. Cambridge: Hackett Publishing Company, 1997), pp. 63-64.

28 Long and Sedley, *The Hellenistic Philosophers*, Vol. 1, p. 163.

29 Plotinus, *The Enneads*, trans. by Steven MacKenna, (New York: Larson Publications, 1992), III.4.3.

meaning, drawing more on the idea of human membership to an ensouled non-hierarchical universe, rather than
the Platonic central theme of hierarchical intellect. In 1957,
Mircea Eliade, complaining of the Platonic notion of cosmos, claimed that 'For him [modern people] the universe
does not properly constitute a cosmos—that is, a living and
articulated unity; it is simply the sum of the material re-
sources and physical energies of the planet'.[30] By this he argued that cosmos was principally the idea of a living unity
and thus he chose to ignore the classical theme of a cosmos based on intellect. This particular version of cosmos
was also articulated by Christopher Chapple who spoke of
cosmology as giving us a place in our universe 'where our
story can be told in such a way that it makes sense to ourselves' and 'the study of the cosmos begins and ends with
the exploration of self'.[31] Thus the modern view of cosmology tends to now include an anthropological outlook which
is focused more on the self and the self's relationship to a
holistic universe. Nicholas Campion acknowledged this
shift when he noted that cosmos now suggests a '"mean-
ing-ful" place in which humanity is an active participant'.[32]
This new definition of cosmos is, however, put to work to
encompass any world view, any ideology concerning the
nature of the world, and thus, from this privileged posi-

30 Mircea Eliade, *The Sacred and the Profane: The Nature of Religion* (1957; repr.
   London: Harcourt, Inc., 1987), pp. 93–94.
31 Christopher Key Chapple, 'Thomas Berry, Buddhism, and the New Cosmology', *Buddhist-Christian Studies* 18 (1998): pp. 147–54, here p. 147.
32 Nicholas Campion, 'Introduction', in *Cosmologies*, ed. by Nicholas Campion
   (Ceredigion, Wales: Sophia Centre Press, 2009), pp. 1–3, here p. 2.

tion, cosmos potentially renders chaosmos invisible.

Hence in order to see chaosmos, one is required to return to Plato's sky-based view of cosmology, that is, the creative force being made up of Reason, both divine and human. Hence this cosmological philosophy is one where the world is governed in a hierarchical manner by a divine logic and, by extension therefore, all of creation can be understood by the use of intellect. Such creation is driven by a clear causal agent, an artisan, be it the divine mind creating the world, or a single human intellect undertaking a creative act, all is ruled by logic. By returning to this earlier view of cosmos, the two states of chaosmos and cosmos become two unique ontological forces, linked and overlapping but at no time ever precluding the other. For it would be wrong to dismiss chaosmos as an expired creative force.

THE ATTRIBUTES OF ASTROLOGY

Astrology assumes as one of its persisting and unchanging tenets that one can read the patterns of the sky and relate these to the dynamic events on earth. Even when the Greeks, most notably the second-century CE polymath Claudius Ptolemy, built a complex and intricate numerically driven astrology, they did not relinquish the notion of Stoic *sumpatheia* but simply allowed their cosmic numbers to 'speak' for it. Ptolemy acknowledged the principle of *sumpatheia* within astrology when he stated 'That a certain power, derived from the aethereal nature, is diffused over and pervades the whole atmosphere of the earth, is clearly evident to all men'.[33] Daryn Lehoux, defined the *sumpatheia* contained in Ptolemy's astrology as the 'co-moving of related bodies across apparent distances' where the related bodies were all things under the orbit of the moon.[34] A hundred years later Plotinus summarised such an all-encompassing sympathetic world as, 'all teems with symbol; the wise man is the man who in any one thing can read another, a process familiar to all of us in not a few examples of everyday experience'.[35] Astrology has never lost its need for the all-linking 'aethereal nature'. Some fifteen hundred years later the French astrologer, Jean-Baptiste Morin (1583–1656) stated this central tenet of astrology by writing, 'There is nothing that is inherent in a man or will be inherent in him that is not signified by the stars in his natal horoscope'.[36] So essential is this attribute that astrology has over its history changed many of its techniques, practices, focus, and attitude to the divine, as it has adapted to differing social, cultural, and political needs, but it is unable to relinquish its need for *sumpatheia*—a sympathetically-linked world where the activity of the heavens is related to life on earth. Astrology is driven by numbers gleaned from the sky, numbers that contain planetary qualities and are given efficacy to describe life on earth through their link with the chaos-

33 Claudius Ptolemy, *The Tetrabiblos* (Mokelumne Hill, CA: Health Research, 1969), Bk. 1.1.2.

34 Daryn Lehoux, 'Tomorrow's News Today: Astrology, Fate, and the Way Out', *Representations* 95 (Summer 2006): pp. 105–22, here p. 108.

35 Plotinus, *The Enneads*, II.3.7.

36 Jean-Baptiste Morin, *Astrologia Gallica Book Twenty-Two Directions*, trans. James Herschel Holden (Tempe, AZ: AFA, 1994), p. 7.

mic notion of *sumpatheia*. In short astrology is the practice of working with cosmic numbers made potent through the notion of *sumpatheia*.

These cosmic numbers are composite in nature as they contain both the quality (symbolism) and quantity (movement) of the planets. Thus, unlike a common integer, these cosmic numbers are mathematical units empowered with planetary meaning and therefore allow for complex themes to be manipulated by mathematics within a horoscope. Casey, cited earlier, had pointed to Plato's reliance on number to describe the divine heavens, to reflect divine reason. Cornford commented on Plato's use of Reason as 'the operation of Reason is carried, so far as may be, into the dark domain of the irrational powers'.[37] The dark domain of the irrational was, for Plato, the world below the orbit of the moon, the world of humankind. This is the world where astrologers pursue Plato's quest for reason for they use these cosmic numbers to explore the shifting vagaries of the sub-lunar realm. Since the time of the Greeks astrologers have never let go of these cosmic numbers and they use these special numbers to generate, layer upon layer, a collection of mathematical techniques that produces a miasma of charts and tables. Today, as in previous times, numbers and their interplay with planetary symbols provide the actual meaning of life for many astrologers.[38] But these cosmic numbers are valueless if they are split from

37 Cornford, *Plato's Cosmology*, p. 210.
38 Bernadette Brady, 'Theories of Fate Among Present-day Astrologers' (PhD Dissertation, University of Wales Trinity Saint David, 2012), p. 289.

the notion of *sumpatheia*.

The order suggested by these cosmic numbers, however, lends itself to allowing both astrologers and critics to forget that these numbers require the existence of *sumpatheia*. For the mathematical flexibility of these numbers gave the impression that astrology belonged solely to the logical, rational cosmos, the world of intellect and predictability. The astrology students in pursuit of numerical accuracy had been grabbling with logarithms for centuries as they struggled to shift earthly time and place into the language of spherical geometry and celestial mechanics. So in 1977 when Commodore introduced its first PC onto the domestic market followed by Apple and Radio Shack personal computers in the same year, astrologers, with their cosmic numbers were eager consumers.[39] Matrix Software, a company founded by the astrologer Michael Erlewine to create and sell astrological software to astrologers, was founded in the same year, 1977, and according to its website, the only other software company still operating that is older than Matrix is Microsoft.[40] This was followed by the founding of Astrolabe Inc. by astrologer Robert Hand and a group of fellow astrologers—Arthur Blackwell, Steve Blake, Gary Christen, and Patricia White—with the release of its first astrological programs for the Radio Shack TRS-80 in 1979.[41] By the mid-to-late 1980s the dominant buyer of advertising in astro-

---

39 http://oldcomputers.net/trs80i.html [Accessed 23 November 2013].

40 http://www.astrologysoftware.com/aboutus.asp [Accessed 8 January 2014].

41 http://alabe.com/history.htm [Accessed 8 January 2014].

logical magazines in the USA and UK were software-related business.[42] Astrologers needed computers and software to handle the heavenly numbers that filled their lives. In my own experience when running a school of astrology with a hundred students in the 1980s and 1990s in South Australia, it was the female astrology student who was introducing the wonders of a personal computer to her household. Understandably, with astrologers themselves enchanted by the computer accuracy of the rational mathematical side of their cosmic numbers, the complex nature of these numbers has tended to be ignored. This has resulted in conflicting and confusing definitions of astrology.

Astrology's nature is that it wears two cloaks, a life lived within *sumpatheia* and wrapped in the surety of numbers. Definitions of astrology, however, tend to focus on only one of these attributes, or at times, neither. In 1995, John Woodruff et al., considered that astrology was,

> a pseudo-science professing to assess peoples' personality traits and to predict events in their lives and future trends in general from aspects of the heavens, in particular the positions of the planets. Astrology is based on ideas which are scientifically unsound and which the great majority of rational people dismiss.[43]

---

42 For examples of this see the *NCGR Journal* (Winter 1986–7) and *The Astrological Journal* 30, no. 6 (1988).

43 John Woodruff, Neil Bone, and Storm Dunlop, *Philips' Astronomy Dictionary: An Illustrated A-Z Guide to the Universe* (London: George Philip Ltd., 1995), p. 20.

Woodruff was suggesting that the presence of number in astrology meant that it should conform to the requirements of science, yet he ignored the chaosmic element of *sumpatheia* required by these numbers to give them efficacy. In contrast, Dion Fortune argued that, 'Ceremonial, and especially talismanic, magic is the essential complement of Astrology; for Astrology is the diagnosis of the trouble, but magic is the treatment of it by means of which the warring forces in our natures are equilibrated'.[44] Here Fortune chose to ignore the number component of astrology and focused on *sumpatheia* defined by her as the spiritual or esoteric side of the subject. In contrast Patrick Curry offered a definition of astrology as '...the practice of relating the heavenly bodies to lives and events on earth, and the tradition that has thus been generated'.[45] Useful and insightful as Curry's definition is, it avoids the actual nature of astrology and focuses instead on its use.

Another trend, as reflected in Curry's definition, has been to take a relative position and define astrology as a product of culture. Campion commented that 'there is indeed a certain postmodern relativism in astrologers' rhetoric', and he then cited his own research which revealed that astrologers selected different astrological techniques based on their convenience rather than holding any intrin-

---

44 Dion Fortune, *The Mystical Qalabah* (1935; repr. London: Ernest Benn, 1972), p. 106.

45 Patrick Curry, 'Astrology', in *The Encyclopaedia of Historians and Historical Writing 2 Vols.*, ed. by Kelly Boyd (London: Fitzroy Dearborn, 1999), pp. 55–57, here p. 55.

sic value in their own right.[46] Astrology can be considered a culturally relative subject dependent on the beliefs of its practitioners. This would explain the subject's shifts and changes and multiplicity of identities. Cultural relativism would also explain astrology's ability to give meaning to life for a diversity of people across cultures and history; nevertheless, to assume that such relativism is the full explanation of the subject would be an error.

Astrology does hold a level of philosophical absolutism which is independent of cultural waves and persists without change across its history. This absolutism is the feature of cosmic number given efficacy through the idea of *sumpatheia*. Astrology, defined by its absolute elements, is an attempt to domesticate the chaosmic phenomenon of *sumpatheia* through the use of cosmic numbers—an attempt by humanity to peer into the void. Any effort to consider the nature of astrology without acknowledging these fundamental and absolute components will suffer from incompleteness. Indeed it may well be that astrology has found itself marooned in a world dominated by cosmos, a world view which can only see the cosmic numbers and has lost the ability to see the essential component of these numbers, chaosmos expressed as *sumpatheia*. Instead of chaosmos, such a world view only sees irrational thoughts in astrology.

This chapter has given a trajectory of the research and

---

46 Nicholas Campion, 'Astrology's place in historical periodisation', in *Astrologies: Plurality and Diversity*, ed. by Nicholas Campion and Liz Greene (Ceredigion, Wales: Sophia Centre Press, 2011), pp. 217–54, here p. 245.

established the parameters of chaosmos and cosmos and the attributes of astrology. The next chapter, chapter two, takes this idea of two unique creative forces, cosmos and chaosmos, and explores them in different creation mythologies. Chapter three then considers the historical path of the privileging of cosmos. Chapter four returns to astrology and looks at the philosophical world view of the time of its creation and its long struggle to be contained in the world defined solely by cosmos. Chapter five discusses the return of chaos with the advancement of Chaos Theory and its expression into the life sciences known as Complexity Science. Chapter six undertakes the mapping of these ideas into the practice of astrology and suggests that astrology is a vernacular expression of complexity and thus draws more from chaosmos than cosmos. The final chapter, chapter seven, provides a discussion on the findings of the research and the final conclusions.

CHAPTER 2:

CREATION MYTHOLOGY, THREE PATHS TO ORDER

THIS CHAPTER CONSIDERS a variety of creation mythologies with the aim of exploring the vernacular view of creation—how human perception has made sense of, and given meaning to, the world. This reveals that the human experience of life senses three modes of creation: one as emergent and polymorphic; another the battle for control and dominance of human gods; and the third based on the assumption of a pre-existing order and the presence of an artisan god. These intuitively derived perceptions of the creation of order are, in a later chapter, revisited and considered in the light of the findings of chaos and complexity science.

Sean Kane describes myths from pre-historical cultures as being stories of patterns, stories which talk of the relationship between plants, animals, and earth.[47] In these stories humanity is not central but is simply one of many players in the emerging patterns of life and from these patterns the gods or goddesses emerged.[48] The images of the divine from this period were predominantly polymorphic blends of human, plant, and animal or half-formed, or shape-changed humans. The *Goddess of Lespugue* found in the Pyrenees Mountains in France (Fig.1) dated some 23,000 years ago is an example of this blending as the female image has a bird-like head with an egg shaped body.[49]

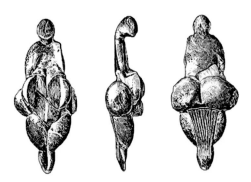

FIGURE 1. 'The Goddess of Lespugue' the Gravettian-Upper Perigordian of about 23,000 BCE found standing on a hearthstone in a shallow cave in the Pyrenees of southern France. The mammoth ivory original is 5 3/4 inches tall, and is now in the Musee L'Homme, France. Image from R. de Saint-Perier (1924).

47 Sean A. Kane, *Wisdom of the Mythtellers* (Ontario: Broadview Press, 1998), p. 36.
48 Kane, *Wisdom of the Mythtellers*, p. 251.
49 R. de Saint-Périer, 'La statuette féminine de Lespugue (Haute-Garonne)', *Bulletin de la Société préhistorique de France* 21 (1924): pp. 81–84.

Such images were also prevalent in Mesopotamia around 1800 BCE as can be seen in *The Queen of the Night* (Fig. 2), which is a female image which has wings and the feet of an eagle. The figure is of a curvaceous naked woman who wears a horned headdress characteristic of a Mesopotamian deity and holds a rod and ring of justice, symbols of her divinity. Her long multi-coloured wings hang downwards and her legs end in the talons of a bird of prey, similar to those of the two owls that flank her. She stands on the backs of two lions, and a scale pattern indicates mountains.

Such images, of which these are just two examples, suggest that the human view of the world was one where life forms and nature were blended, interlinked or interwoven. As Henri Frankfort stated, in discussing the period of the Old Kingdom in Egypt, the view of the divine could be a blend of any life form as well as a mountain or the wind.[50]

ORDER CREATED BY EMERGENCE

This was a time before humanity dared to make the gods in its own image. This blending was embedded in the philosophy of creation from chaos. Chaos as a source of creation was, according to Merlin Stone, reflected in the earliest known creation myth which is from the Semitic people of Mesopotamia around 5,000 BCE:

---

50 Henri Frankfort, *Ancient Egyptian Religion* (New York: Columbia University Press, 1948), p. 9.

Queen of Heaven, Goddess of the Universe, the One who
walked in terrible chaos and brought life by the law of love
and out of chaos brought us harmony and from chaos She
has led us by the hand...[51]

Chaos, as a creative force, was also written in the Pyramid
Text of around 2,200 BCE where the Heliopolitan priests
described the concept of creation coming out of chaos.[52]
These texts talked of the chaotic void, Nun, described by
Francoise Dunand and Christiane Zivie-Coche as that
which 'pre-existed, the uncreated, the unformed, the un-
differentiated, the atemporal, to which was opposed, after
its creation, the cosmos, which was ruled by order'.[53] Nevertheless Dunand and Zivie-Coche point out that for the
Egyptians, Nun was unchanged after the creation of cosmos.

It was from the void of Nun, metaphorically seen as the
waters of the Nile, that the Egyptian creator god emerged,
one expression of which was the ram-headed Khnum (Fig.
3). Khnum's emergence was known as the 'First Occasion'
and, in the form of a potter Khnum began to create the rest
of life, dipping his/her hand or pot into the patterns forming in the silt of the Nile and shaping these patterns into life
on a potter's wheel. Khnum was a polymorphic water deity
who was called Father of Fathers and Mother of Mothers.

51 Merlin Stone, *Ancient Mirrors of Womanhood: Our Goddess and Heroine Heritage. Vol. 1* (New York: New Sibylline Books, 1979), p. 107.

52 F. Guirand, *Egyptian Mythology* (New York: Tudor, 1965), p. 27.

53 Francoise Dunand and Christiane Zivie-Coche, *Gods and Men in Egypt,
   3000 BCE to 395 CE* (London: Cornell University Press, 2004), p. 46.

FIGURE 2. 'The Queen of the Night' 1800–1750 BCE. Old Babylonian, from southern Iraq. This plaque, height 49.5 cm, width 37.00 cm, and thickness around 4.8 cm, is made of baked straw-tempered clay, modelled in high relief. It is held in the British Museum. Image courtesy of the British Museum.

Periodically Khnum would dissolve back into the void only to re-emerge again. This was known as a 'return to the first occasion' a time of remaking.[54] This was a cyclic creation myth, but like chaosmos discussed in the previous chapter, this cyclic nature did not reproduce itself identically.

FIGURE 3. The ram-headed Khnum, the potter. Khnum is a polymorphic deity known as Father of Fathers and Mother of Mothers and is considered to be one of the earliest Egyptian deities established in the Predynastic Period (5464–3414 BCE). He emerged from the void and then proceeded to create the rest of life on his potter's wheel. Image Wikimedia Commons.

This Egyptian view was not unique for, as Michael Bütz stated, 'chaos was an essential, even pivotal, concept with

54 Dunand and Zivie-Coche, *Gods and Men in Egypt*, pp. 46–51.

the ancient Taoists, Egyptians and Native Americans at least a thousand years before Greek civilization imagined it as a philosophical form'.[55] Bütz found that in these myths chaos was an essential state for the creation of new forms and that themes such as a conscious god/goddess who represents chaos, or a god or several gods needing to return to something like chaos to create, were prevalent throughout many of the creation myths.[56] Kane pointed out that such myths contained themes of darkness, void, and the emergence of half-life, half patterns that keep returning to the darkness to reform, reshape, and create another pattern.[57] One such cyclic creation myth comes from native Australians:

> In the beginning the earth was a bare plain. All was dark. There was no life, no death. The sun, the moon, and the stars slept beneath the earth. All the eternal ancestors slept there, too, until at last they woke themselves out of their own eternity and broke through to the surface. When the eternal ancestors arose, in the Dreamtime, they wandered the earth, sometimes in animal form—as kangaroos, or emus, or lizards—sometimes in human shape, sometimes part animal and human, sometimes as part human and plant.
>
> Two such beings, self-created out of nothing, were the Ungambikula. Wandering the world, they found half-made

55 Michael Bütz, *Chaos and Complexity—Implications for Psychological Theory and Practice* (Washington, DC: Taylor and Francis, 1997), p. 206.

56 Bütz, *Chaos and Complexity*, pp. 207–9.

57 Kane, *Wisdom of the Mythtellers*, p. 65.

human beings. They were made of animals and plants, but were shapeless bundles, lying higgledy-piggledy, near where water holes and salt lakes could be created. The people were all doubled over into balls, vague and unfinished, without limbs or features. With their great stone knives, the Ungambikula carved heads, bodies, legs, and arms out of the bundles. They made the faces, and the hands and feet. At last the human beings were finished.[58]

Another example comes from China where chaos slowly gave rise to the creation of the world:

In the beginning, the heavens and earth were still one and all was chaos. The universe was like a big black egg, carrying Pan Gu inside itself. After 18 thousand years Pan Gu woke from a long sleep. He felt suffocated, so he took up a broad-axe and wielded it with all his might to crack open the egg. The light, clear part of it floated up and formed the heavens, the cold, turbid matter stayed below to form earth. Pan Gu stood in the middle, his head touching the sky, his feet planted on the earth. The heavens and the earth began to grow at a rate of ten feet per day, and Pan Gu grew along with them. After another 18 thousand years, the sky was higher, the earth thicker, and Pan Gu stood between them like a pillar 9 million li in height so that they would never join again.[59]

58 Australian Myths, http://www.dreamscape.com/morgana/miranda.htm#AUS [Accessed 30 September 2004].
59 Chinese Myths, http://www.dreamscape.com/morgana/ariel.htm#HAW [Accessed 30 September 2004].

The myth describes the death of Pan Gu and his decaying body which then becomes the *prima materia* of life. It is reasonable to claim that the majority of creation myths from across cultures contained this essence of creation or life emerging from the void of chaos. This mythological chaos was a place of potential, a place or state from which life emerged, again and again and again.[60] Eliade claimed that humanity developed these myths as a dialogue with these forces, a two-way or multi-layered exchange of one part of the web influencing the other.[61] In such creation mythology one can see the fingerprint of chaosmos with its cyclic, scale-shifting, emergent manner of creating order.

CHAOSMOS, *SUMPATHEIA* AND OMENS

The nature of a culture's creation mythology will, in fact, inform an individual of what is and what is not possible. It will define the culture's notion of rational thinking. With a creation mythology based in chaosmos the world becomes an interlinked web of relationship and is embraced by the Stoic idea of *sumpatheia*. In such a world superstitions and omens are a rational concept. It is rational, and consistent with the beliefs about how things are created that a small

---

60 See J. Briggs and F. D. Peat, *The Turbulent Mirror: An Illustrated Guide to Chaos Theory and the Science of Wholeness* (New York: Harper & Row, 1989); James Gleick, *Chaos: Making a New Science* (New York: Viking-Penguin, 1987).

61 Eliade, *The Sacred and the Profane*, pp. 32–33; Kane, *Wisdom of the Mythtellers*, p. 14.

action intentionally or accidentally undertaken will have larger implications in another area of life, independent of any measurable or visible causal link. This gives rise to omens.

Omens are made up of a pair of events and are defined by Erica Reiner as the protasis (if-cause) and the apodosis (the forecast) and both are indifferent to scale: the size or nature of the protasis (a mad dog, a cat, a cloud passing overhead, or a celestial event) was not relevant to the size of the expected apodosis (health of the king, the fertility of the herd, the weather).[62] For example, Reiner gives a translation of some Babylonian Venus omens from about the seventh century BCE: 'If Venus's position is green: pregnant women will die with their foetuses—Saturn stands with her'.[63] This is a celestial omen linking a visual event in the sky to a human event on earth with no regard to the size of the planet Venus in relationship to the size of a pregnant woman. Furthermore, Reiner also pointed out that if the protasis was repeated, then it was expected that the associated apodosis would also occur in a similar manner.

Thus for the Babylonian culture the adoption of a chaosmic creation myth provided a form of 'physics' which by extension produced a definition of what was rational, logical, and even intellectual, and which 'logically' embraced the notion of omens and superstition.

---

62 Erica Reiner, 'Babylonian Celestial Divination', in *Ancient Astronomy and Celestial Divination*, ed. by N. M. Swerdlow (London: MIT Press, 1999), pp. 21–37, here p. 23.

63 Reiner, 'Babylonian Celestial Divination', p. 33.

ORDER CREATED OUT OF CONFLICT

Nevertheless, chaosmos provides only one type of creation myth. Norman Cohn in talking of ancient faiths, points out that there was a natural polarity between chaosmos and cosmos.[64] In his version of creation, the *Timaeus*, Plato exemplified cosmos as, 'Now everything that comes to be must of necessity come to be by the agency of some causes', thereby implying that cosmos is built on a measurable antecedent causality.[65] This view of preferred order was in juxtaposition with the nature of chaosmos, and such juxtaposition produced a tension between *Kosmos* and chaosmos with the promise of surety as the prize offered by *Kosmos*. Philip Kuberski argued that *Kosmos* was and still is pleasing, re-assuring, and confirming, guaranteeing the surety of one day to the next, offering humanity knowledge and order, while Chaos was feared, as it was the place where one could be swallowed up, where the power of the intellect could be lost, and where the boundaries of the self dissolved.[66] Eliade argued that this fear could produce the idea of conflict as the features of chaos challenged the existence of the ordered cosmos.[67] When such conflict occurred in creation myths Eliade defined them as 'blood-drenched cosmologies', for they underpinned the notion that cre-

---

64 Norman Cohn, *Cosmos, Chaos, and the World to Come*, 2nd ed., (New Haven, CT: Yale Nota Bene, 2001), p. 3.

65 Plato, *Timaeus*, 28a.

66 Kuberski, *Chaosmos: Literature, Science, and Theory*, p. 37.

67 Eliade, *The Sacred and the Profane*, p. 22.

ation required violence.[68]

Ralph Abraham discussed one such conflict creation myth, the Sumerian hymn known as the *Enuma Elish* the earliest written version dating from the twelfth century BCE. In this myth Enlil's (earth) son Niurta defeats the dragon or serpent of chaos, Basmu. Chaos is also personified as the dragon Tiamat:

> When above unnamed was the heaven,
> The earth below by a name was uncalled,
> The primeval deep was their begetter,
> The chaos of Tiamat was the mother of them all.[69]

Tiamat engaged in a struggle with Marduk, the later replacement of the Sumerian deity of Earth's son (Niurta), and eventually Marduk acquired new weapons and overthrew Tiamat (chaos) and split her into two. One part he left in the earth and the other he placed above the earth to make the covering of the heavens. This is, to use Eliade's term, a blood-drenched cosmology. This particular myth has had long cultural legs, for the battle of Marduk against the chaotic serpent Tiamat was later reflected in *Revelation* 12:7–9 with the battle between the archangels of Michael and Lucifer—Satan:

> And there was war in heaven: Michael and his angels fought against the dragon; and the dragon fought his an-

68 Eliade, *The Sacred and the Profane*, p. 51.
69 R. Abraham, *Chaos, Gaia, Eros* (New York: HarperSanFrancisco, 1994), p. 128.

gels. And prevailed not; neither was their place found any more in heaven. And the great dragon was cast out, that old serpent, called the Devil and Satan, which deceiveth the whole world: he was cast out into the earth, and his angels were cast out with him.[70]

Such a conflict was also repeated in the fourth century CE story of Saint George and his victory for Christianity over the pagan religions symbolised by his killing of a dragon.[71] In such myths the dragon, with its chaotic symbolism, was seen to reflect all that was negative and destructive and was to be condemned, overthrown, damned, killed, beheaded, and eliminated for all time. These battles were not ones in which resolutions or compromises were considered. They demanded total victory and the total eradication of all of the dragon's elements. Abraham argued that the west, historically and culturally, has been in a continuum of this ancient battle which seeks the permanent removal of the dragon of chaos and the annihilation of chaosmos.[72] In discussing his own field of psychology and the use of chaotic concepts in his profession, Bütz supported Abraham's idea and argued that this battle was also the force behind the Christian missionary movements which Bütz viewed as the desire to rid the world of chaos by the conversion of the primitive or evil savages, '...it appears Christians condemn

70 *The Dartmouth Bible*, ed. by Roy B. Chamberlin and Herman Feldman (Boston, MA: Houghton Mifflin, 1950).

71 Dorothy Margaret Stuart, *Book of Chivalry and Romance* (London: G. G. Harrap, 1933).

72 Abraham, *Chaos, Gaia, Eros*, pp. 189–98.

chaos as a thing associated with the devil and therefore, to discuss disorderly ideas would surely be of the devil'.[73]

This battle was also expressed in popular culture through Ridley Scott's film *Alien* (1979), a modern box-office success and now of cult standing. In the film the evil female monster was in conflict with a female of the realm of cosmos played by Sigourney Weaver. These two were locked in mortal combat. H. R. Giger's famous female monster wanted to breed with the human race and therefore, from cosmos' point of view, potentially destroy all of humanity. This fertile, salvia-dripping, scaly, female monster can be likened to a version of Tiamat. Weaver's task in Scott's film was not to negotiate with the monster but to totally eliminate her from the face of the universe.

These are creation myths fighting for the order of cosmos where what is defined as rational is all that can be framed within a clear linear causality. They are also creation myths that require conflict and as such can fuel nations to go to war to colonise and spread their preferred culture. The 'Marduks' of the culture will seek to eradicate what they define as irrational, primitive, and illusional thinking which are considered chaotic elements existing in their society which can potentially infect or consume the whole world. Thus the continual presence of astrology in contemporary culture still attracts the wrath of the 'Marduks' of cosmos.

73 Bütz, *Chaos and Complexity*, p. 214.

ORDER CREATED OUT OF INTELLECT

Separate from chaosmos and mythologies which are in conflict with chaosmos, there are creation myths based in intellect which entirely ignore the existence of chaos. For example, Plato produced his number-driven cosmic creation myth with an intellectual artisan god, the demiurge, at its centre.[74] This clear logical being, an artisan-God, is also present within Christian creation mythology, as stated in the Bible:

> In the beginning God created the heavens and the earth. The earth was without form and void and darkness was upon the face of the deep; and the Spirit of God was moving over the face of the waters.
>
> And God said, 'Let there be light' and there was light, And God saw that the light was good; and God separated the light from the darkness. (*Genesis* 1:1-4)[75]

In this Biblical creation story order pre-existed and was known as God, and one action logically followed another with each action providing the basis or cause for the next. In the beginning God created light by separating it from the darkness. Then He created evening from which we had the first dawn. Having created the idea of day, God separated sky from earth and created the heavens. On the second day He separated the waters of the earth to create land and cre-

---

74 Plato, *Timaeus*, 40c.
75 *The Dartmouth Bible.*

ated the plant kingdom to grow upon the dry land. On the third day He created the sun and the moon and the starry sky to act as signs for the seasons. On the fourth day God created life in the seas, and on the fifth day He created life on dry land. On the sixth day God created man in His image and gave him rulership over all of his creations. On the seventh day He rested, giving to His new-found order the concept of a holy day. There is, however, no eighth day as there are no cycles in this linear story.

This myth produced creation through a single mind with a clear plan. The plan was implemented methodically step by step, and as each step was implemented, this mind placed a value judgment on His work. From the beginning of this creation myth, the creative force of order announced that the light was good and, by inference, that the dark was bad and to be avoided. It was a powerful story because it dictated the way in which things happen in the world made by an artisan-God. It told us that the world was created in a logical causal manner and all new order which was created in this world would therefore also be produced in a logical linear causal manner. This style of creation myth, like the cosmic conflict myths, informed its culture that that which was rational could only be that which had a known antecedent linear cause. Any suggestion supporting any other form of creative force was at best considered simply an illusion or at worst heretical. Order was the norm and chaos was viewed as a disorganised place waiting to be ordered.

This chapter has considered the three ways that mythology described the creation of order and the resulting definitions of rational thought. Order from chaotic creation mythology was emergent, polymorphic, and non-anthro-pological in orientation and based in cycles. Rational think-ing based in these myths included omens and superstitions. In contrast, cosmological myths create order in two ways, by entering into conflict with the chaotic forces, or by actually denying the existence of a creative chaos. The chaotic myths required that order needed to be guarded and maintained, as discussed in the previous chapter through the Assyrian astrologer/priest and their need to watch the heavens for signs of disorder. The cosmic myths, however, begin with pre-existing order as a divine intellect, and this order was personified as the artisan god, crafting the world using a linear dynamic methodology. Using *Genesis* as the key example this myth contained no cyclic elements and no threat of the loss of order. Order being championed by God could be assumed and was assured. Rational thought was defined by these two types of cosmic creation myths as being that which was underpinned by antecedent linear causality. The next chapter looks at how the idea of creation from cosmic order came to dominate the modern world view.

CHAPTER 3:

THE PRIVILEGING OF COSMOS

THIS CHAPTER CONSIDERS the historical journey that lead to the privileging of cosmos and the eventual exclusion or denial of the ontological force of chaosmos. It argues that although cosmos' victory has been largely successful it has, however, left a vacuum between the supposed ways the world is meant to work versus the actual lived experience. Such a vacuum has created, for many individuals, an existence where the dominant world view of cosmos suggests that their experience of life, one filled with coincidences, synchronicities and re-occurring patterns, is actually meaningless and any other belief is illusional. In conclusion I argue that the continual existence of such illusional ideas in society suggests that chaosmos is present as an active creative ontological force.

THE RISE OF COSMOS

The shift from the creative void of chaosmos to the divinity of ordered cosmos was not a linear transformation. The creative force hidden in the darkness or turbulence of chaos was a theme central to the philosophy of Heraclitus (540–480 BCE) when he proposed a law that he called *logos*, where forces moved towards each other. Klaus Mainzer paraphrasing Heraclitus stated, 'What is opposite strives towards union, out of the diverse there arises the most beautiful harmony, and the struggle makes everything come about in this way'.[76] But the flux or *logos* of Heraclitus had philosophical problems for the Greek quest for understanding. If everything was in a state of flux, then everything was constantly changing, and if everything was changing, then nothing could be known. As Heraclitus wrote, 'The river where you set your foot just now is gone— those waters giving way to this, now this'.[77] The creative chaos, argued Heraclitus, did not allow for the concept of knowledge, yet such a philosophical terminus was unacceptable to the needs of those who sought a stable, and thus knowable, world.

Parmenides of Elea (c. 500 BCE) presented a counter-argument to Heraclitus and argued that the world was solid and uniform without motion and time.[78] For Parmenides,

76 Klaus Mainzer, *Thinking in Complexity: The Complex Dynamics of Matter, Mind and Mankind* (London: Springer-Verlag, 1994), p. 17.

77 Heraclitus, *Heraclitus Fragments* (New York: Penguin Classics, 2003), fg. 41.

78 Mainzer, *Thinking in Complexity*, p. 18.

this Eleatic philosophy of unchanging 'being' yielded a singular end state of the highest symmetry. Plato also appeared to answer Heraclitus by suggesting that the ever-changing flux could exist in the land of the living, the sub-lunar world, but that the divine or godly world was perfect, orderly, totally stable and consistent. In the *Republic*, Plato was critical of learning by observation, for the physical senses could lead one towards chaos and uncertainty; only the intellect could produce truths:

> Then if, by really taking part in astronomy, we're to make the naturally intelligent part of the soul useful instead of useless, let's study astronomy by means of problems, as we do geometry, and leave the things in the sky alone.[79]

J. V. Luce argued that Plato considered that there was nothing that belonged to the body, or the visible, the sensual, or phenomenological which could be used as a source of truth.[80] To allow for the existence of knowledge therefore Plato created two worlds. One was the place of 'being', the perfect, stable and orderly world which could be known if one applied Reason. This world belonged to the divine. The other was that of 'becoming', the sub-lunar world where all things were subject to change.[81] With knowledge existing only in the orderly divine world, the measure of the verity

79 Plato, *Republic*, 530b.
80 J. V. Luce, *An Introduction to Greek Philosophy* (London: Thames and Hudson Ltd., 1992), p. 99.
81 Plato, *Timaeus*, 28a.

of a 'truth' was linked to its measure of elegant simplicity or beauty. According to Bernard Williams, Plato emphasised this split across most of his works, with god as knowledge, and earth and thus the body as the unknowable.[82] Mainzer made a similar argument stating that Plato's attitude to the retrograde movement of the planets and his desire to explain them within his perfect circular model had implications for the development of ideas and research up to the present day. It was Apollonius of Perga (c. 210 BCE) who Mainzer claims eventually answered Plato's question by suggesting that the planetary orbits contained epicycles: smaller perfect circles located at points on bigger perfect circles.[83]

In response to the perfect world of Plato, Aristotle (384–322 BCE) introduced the ideas of Form and Matter as a way to bringing order to the ever-changing shape of life. Luce argued that Form was that which made the 'thing' what it was, while the Matter was the substance that had been shaped by the Form.[84] Aristotle (*Physics* II.3) also suggested that Matter had a potential for being formed and it was not until Matter has been formed does reality come into being.[85] Unlike Plato, Aristotle believed the physical realm

---

82 Bernard Williams, 'Plato—The Invention of Philosophy', in *The Great Philosophers*, ed. by Ray Monk and Frederic Raphael (London: Phoenix, 2001), pp. 47–92, here p. 88.

83 Mainzer, *Thinking in Complexity*, pp. 24–25.

84 Luce, *An Introduction to Greek Philosophy*, p. 116.

85 These ideas are represented throughout Aristotle's work, notably in *Physics* II.3, and in *Metaph.* A.3 ff. See also *Part. An.* 639b12ff, *APo.* II.11; *Metaph.* D.2; *Gen. et Corr.* 335a28–336a12.

was not imperfect and concluded that the physical realm, despite always being in flux, moved towards specific ends, demonstrating teleology (*telos*='end' or 'purpose'). The essence of a thing, he argued, did not lie outside the physical realm but was contained within it.[86] Notwithstanding the much-debated concept of Aristotle's *telos*, Anthony Mansueto argued that Aristotle's work was blended with Plato's ideas of a perfect world to yield a research approach to the natural world. This approach considered that the natural world could be understood through observation, but with the understanding that all answers or reasons concerning the physical sub-lunar world were superior or closer to the 'truth' if they were simple and elegant.[87] As Ilya Prigogine and Isabelle Stengers noted, the greater the elegance of the answer, the closer the answer was to the divine; the closer the answer was to the divine, the closer it was to *the truth*.[88]

Thus Nicolaus Copernicus (1473–1543) strove to find a solution to the planetary motion which fitted the elegant and simplistic framework laid down by Plato. Looking for an elegant solution to the observed retrograde movement of the planets, he placed the sun in the centre and allowed the earth to orbit on one of the perfect circles. He therefore explained the retrograde motion of the planets in a simpler

---

86 See for example, Aristotle, *Phys.* 194b13; *Metaph.* 1032a25, 1033b32, 1049b25, 1070a8, 1092a16.

87 Anthony Mansueto, 'Cosmic Teleology and the Crisis of the Sciences', *Philosophy of Science* (1998) http://www.bu.edu/wcp/Papers/Scie/ScieMans.htm [Accessed 29 September 2004].

88 Ilya Prigogine and Isabelle Stengers, *Order Out of Chaos: Man's New Dialogue with Nature* (New York: Bantam Books, 1984), p. 7.

fashion. As Mainzer comments, 'Simplicity was not only understood as the demand for an economical methodology, but, still for Copernicus, as a feature of truth. Thus, the astronomical doctrine from Plato to Copernicus proclaimed: reduce the apparent complexity of the celestial system to the simple scheme of some true motions'.[89] Although sitting completely within neo-Platonism, Copernicus' new model of the solar system began a one-hundred-year period of change where many different models of the solar system were proposed. The resultant upheaval provided the intellectual space for the work of Johannes Kepler (1571–1630) and Galileo Galilei (1564–1642).

Kepler's observation of the supernova of 1604, presented in his work titled *De Stella Nova* (published in Prague, 1606), was more than just an interesting observation in astronomy. It was the recognition of change in the perfect unchangeable divine realm. Earlier in his *Mysterium cosmographicum* (published in Tübingen, 1596), Kepler had pursued Plato's thinking with regard to perfect shapes where simplicity was understood as a harbinger of truth. With Plato's perfection in mind, he created a model of the solar system that based planetary distance on Plato's regular solids. However, in his later work *Astronomia Nova* published in 1609, he abandoned the Platonic concepts of simplicity.[90] In this work Kepler produced his first two laws of planetary

---

89 Mainzer, *Thinking in Complexity*, p. 29.

90 Dedre Gentner et al., 'Analogical Reasoning and Conceptual Change: A Case Study of Johannes Kepler', *The Journal of the Learning Sciences* 6 (1997): pp. 3–40, here pp.12–14.

motion: the first law stating that the planets moved in ellipses with the sun at one foci of the ellipse, and the second law stating that the arc described by the orbiting planet describes equal area in equal time.[91] Hence the planets speeded up and slowed down depending on their position on the ellipse of their orbit (Fig. 4).

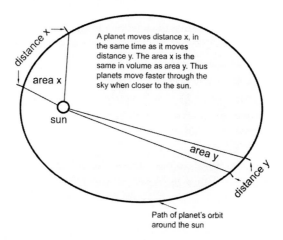

A planet moves distance x, in the same time as it moves distance y. The area x is the same in volume as area y. Thus planets move faster through the sky when closer to the sun.

distance x

area x

sun

area y

distance y

Path of planet's orbit around the sun

FIGURE 4. Kepler's second law of planetary motion described the speed of a planet travelling in an elliptical orbit around the sun. It stated that a line between the sun and the planet swept equal areas in equal times. Thus, the speed of the planet increased as it neared the sun and decreased as it receded from the sun. Image by author.

91 Johannes Kepler, *Astronomia Nova* (1609; repr. London: Cambridge University Press, 1992).

Kepler had removed the perfect circles from the heavens, and, combined with the observation of the supernova of 1604, the sky was now a place of change. In 1609, Kepler summarised this shift of attitude to the heavens in a letter to Herwart von Hohenzollern, the Catholic Chancellor of Bavaria, in which he wrote of his desire to return order to the heavens but under new rules:

> My aim is to show that the heavenly machine is not a kind of divine, live being, but a kind of clockwork, insofar as nearly all the manifold motions are caused by a most simple, magnetic, and material force, just as all motions of the clock are caused by a simple weight. And I also show how these physical causes are to be given numerical and geometrical expression.[92]

In the period when Kepler was seeking to replace the divine heavens with a mechanical order, the philosophical approach to the sub-lunar world was also challenged. Galileo, a contemporary of Kepler, was interested in motion in the sub-lunar world. In his *Dialogue on the Two Chief World Systems*, Galileo used a dialogue between Simplicio and Sagredo to show his attitude to Plato's consideration that truth only existed in the super-lunar realms:

> Simplicio: But I still say, with Aristotle, that in physical (*naturali*) matters one need not always require a math-

92 Arthur Koestler, *The Sleepwalkers: A History of Man's Changing Vision of the Universe* (Harmondsworth, UK: Penguin Books Ltd., 1959), p. 345.

ematical demonstration.

Sagredo: Granted, where none is to be had; but when there is one at hand, why do you not wish to use it?[93]

At this time the organic, sub-lunar world could only be understood through philosophy. Yet Galileo proceeded to measure, weigh, and find laws, which enabled him to explain and predict some of the workings of the sub-lunar world.[94] His approach to the sub-lunar world was so radical that the French philosopher René Descartes (1596–1650) rejected Galileo's physics totally because, rather than investigate the cause of motion or heaviness, Galileo had instead simply measured it.[95]

Although Kepler and Galileo corresponded with each other, both pursued their own endeavours: one, motion of the heavens; the other, motion in the sub-lunar world. Nevertheless their work overlapped; as Kepler caused order to fall from the sky, so it found a home in the sub-lunar world through the work of Galileo. Eventually cosmic thinking reclaimed its rulership over the sky, but for a short time it appeared as if Tiamat had returned to the heavens. This turbulence was reflected in the poem of John Donne written in 1611, *The First Anniversary*, when he wrote:

93 Galileo Galilei, *Dialogue Concerning the Two Chief World Systems* (1632; repr. New York: The Modern Library, 2001), p. 14.

94 Leonardo Olschki, 'Galileo's Philosophy of Science', *The Philosophical Review* 52 (1943): pp. 349–65, here p. 351.

95 Stillman Drake, *Galileo* (Oxford: Oxford University Press, 1996), p. 11.

And new Philosophy calls all in doubt,
The Element of fire is quite put out;
The sun is lost, and th'earth, and no man's wit
Can well direct him, where to look for it.
And freely men confess, that this world's spent,
When in the Planets, and the Firmament
They seek so many new; they see that this
Is crumbled out again to his Atomis.
'Tis all in pieces, all coherence gone;
All just supply, and all relation:
Princes, subject, father, son, are things forgot...

Donne, like Geoffrey of Monmouth cited in chapter one, turned to the sky to talk of chaos, but this time with good reason, for Kepler had disrupted the sky. Nevertheless Kepler eventually did succeed in his new view of the heavens. He discovered his third law of planetary motion—the square of the orbital period of a planet is proportional to the cube of the semi-major axis of its orbit. He published this law in *Harmonices mundi libri V* (Linz, 1619) thus revealing that he had found the 'elegant' relationship between the orbital period of a planet and its distance from the sun. This returned the surety of simplicity and order to the heavens. A generation later, Isaac Newton (1643–1727) also helped this domination of cosmic order with his mathematical development of differential equations, the equations which deal with shifting variables and are the foundations of calculus (*see* Glossary). Indeed after Newton, Fritjof Capra asserted that, 'Newton's differential equations became the mathematical foundation of the mechanistic paradigm.

The Newtonian world machine was seen as being completely causal and deterministic'.[96] The idea, therefore, of chaos as a creative force no longer held dominion over earth or sky and was dismissed as belonging to the 'primitive' beliefs of superstition and ignorance.

Followers of science by this time considered that all things could be known. Marquis Pierre Simon de Laplace (1749-1827), the French mathematician and astronomer, summarised this position by stating:

> We may regard the present state of the universe as the effect of its past and the cause of its future. An intellect which at any given moment knew all of the forces that animate nature and the mutual positions of the beings that compose it, if this intellect were vast enough to submit the data to analysis, could condense into a single formula the movement of the greatest bodies of the universe and that of the lightest atom; for such an intellect nothing could be uncertain and the future just like the past would be present before its eyes.[97]

There was, of course, disagreement with this mechanical, knowable, ordered world notably in the work of the philosopher Immanuel Kant (1724-1804) who suggested that an organism 'cannot only be a machine, because a machine

---

96 Fritjof Capra, *The Web of Life: A New Scientific Understanding of Living Systems* (New York: Doubleday, 1996), p. 119.

97 Pierre-Simon Laplace, *A Philosophical Essay on Probabilities* (1812; repr. New York: Cosmio, 2007), p. 4.

has only moving force; but an organism has an organising force...which cannot be explained by mechanical motion alone'.[98] Here Kant was describing a feature of chaosmos, that life contained a self-organising force. From the work of Kant, along with that of Johann Wolfgang von Goethe (1749–1832) and Karl Wilhelm Friedrich von Schlegel (1772–1829), a romantic philosophy arose in Germany which supported chaotic principles of creation. Schlegel voiced this philosophy when he advocated that although knowledge and the advance of knowledge by reductionism was a productive endeavour, the greater noble cause was the inner mental life (*geistiges Leben*), which could not be reduced in this way.[99]

The work of Charles Darwin (1809–1882), however, silenced the debate. Darwin's publication of his theory of evolution was held as proof of the validity of the reductionist, mechanised world view. As Edward Larson claimed, Darwin's theory attempted to show that complex organisms could evolve purely by the pressure of survival of the fittest, thereby removing the need for any teleological or natural self-organising elements.[100] Stuart Kauffman argued that from this time forth life was viewed as solely

---

98 Immanuel Kant, *Kritik der Urteilskraft*, ed. by G. Lehmann (1781; repr. Stuttgart: Reclam, 1971), p. 340.

99 Friedrich von Schlegel, *The Philosophy of Life and Philosophy of Language in a Course of Lectures* (London: Henry G. Bohn, 1847), p. 4.

100 Edward J. Larson, *Evolution: The Remarkable History of a Scientific Theory* (New York: Modern Library, 2004), pp. 79–111.

based on chance and was seen as mechanical.[101] With the advent of Darwin's work it appeared as if Marduk had finally defeated the dragon of chaos, Tiamat. All things could now be known. Yet in spite of the domination of order through cosmos and the denigration or dismissal of chaosmos, the creative chaos never left. The perception of creation coming from the chaotic void is the hidden force behind the human concepts of superstitions, folklore, 'old wives' tales' or what James Reston called the 'hocus pocus' of the common folk.[102]

CHAOSMOS AND THE 'HOCUS POCUS' OF THE COMMON FOLK

Omens and superstitions still play a role in many people's lives. Colin Campbell claimed that the, 'widespread continued presence of superstitious belief and practice in modern industrial societies present sociologist with a problem'.[103] Campbell argued that such beliefs were problematic as they were non-rational beliefs which should 'fade away in the face of that secular rationalism and empiricism'.[104] In light of the arguments of chapter two, such beliefs of

---

101 Stuart Kauffman, *At Home in the Universe: The Search for the Laws of Self-Organization and Complexity* (New York: Oxford University Press, 1995), p. 6.

102 James Reston, Jr., *Galileo: A Life* (New York: Harper Collins, 1994), p. 13.

103 Colin Campbell, 'Half-belief and the Paradox of Ritual Instrumental Activism: A Theory of Modern Superstition', *The British Journal of Sociology* 47 (1996): pp. 151–66, here p. 151.

104 Campbell, 'Half-belief and the Paradox of Ritual Instrumental Activism', p. 151.

course are deemed 'non-rational' as they are not supported by the dominant creation mythology which considered that an intellectually derived order is the *only* creative ontology. Additionally, Campbell claimed that levels of superstition do not actually drop as society modernises, rather the type of the superstitions change from one generation to another.[105] Campbell's claim of unchanging levels of superstition was supported by the work of Edmund Conklin, who in 1919 measured the presence of superstitions amongst US college students, and he found that 82% of the 457 students he surveyed admitted to holding some superstitious beliefs.[106] In the face of this type of evidence which revealed the constant persistence of these 'non-rational' beliefs in society, Campbell concluded that superstition was a 'continuing feature of human experience' and played a part in traditional as well as modern societies.[107]

Such superstitions can be quite public, particularly when existing in the sporting arena. The Australian cricketer Steve Waugh (captain 1999–2004) set new world records in batting, but he never went out to bat without his old red handkerchief tucked in his pocket partly showing against his cricket whites.[108] He believed that by carrying the lucky red handkerchief, he would have a better chance

---

105 Campbell, 'Half-belief and the Paradox of Ritual Instrumental Activism', p. 153.
106 Edmund S. Conklin, 'Superstitious Belief and Practice Among College Students', *The American Journal of Psychology* 30 (1919): pp. 83–102, here p. 87.
107 Campbell, 'Half-belief and the Paradox of Ritual Instrumental Activism', p. 152.
108 Personal knowledge from commentaries on the ABC (Australian Broadcasting Commission) while they broadcasted live coverage of cricket matches.

of repeating his high scoring performance, although he did not expect to always score the exact same number of runs. Thus for Waugh the protasis (the if-cause as discussed in chapter two) was his red handkerchief and the apodosis (the forecast) was the hope of gaining a high score. At first glance one is tempted to dismiss this as personal idiosyncratic behaviour, but on closer inspection what is revealed is that Waugh's personal beliefs around creation mythology were not in accord with his society's supposed dominant views on creation. Waugh's actions indicated that he subscribed, at least in part, to a chaosmos creation of order. Such a creation of order is emergent where cycles and events repeat but not in an identical manner to their earlier expression. Additionally, such emergentism exists across scale, thus a red handkerchief carried in the pocket on the day of a good batting score can be used as an attempt to 'return to the first occasion' again and again, in the hope or belief that a high batting score will follow. This is a personal ritual which may have been instrumental, at least at a psychological level, to help Waugh's batting skills. Superstition is the repetition of a behaviour which is considered, by the performer, to be 'causally' linked to a favoured outcome regardless of scale or form. Waugh of course is only one example of such superstitions within the professional sporting arena.

The existence of superstition across cultures is just one anomaly in modern life. There are many others that are deemed by rationalists as illusional. The un-defined notion of a fated life, where it would seem that an individual was destined to arrive at a certain situation, any form of divination, any notion of luck or being unlucky, placing meaning on co-incidences, looking for omens or signs to give understanding of an event, are all common everyday accepted behaviours. The human propensity for such beliefs, in spite of their condemnation, offers evidence of the existence and presence of chaosmos in the experience of a lived life.

This chapter has explored the historical journey of the privileging of cosmos over the ontological force of chaosmos. It has also commented on the consequences of this privileging which sets up a two level style of life for many individuals where they have to live with their behaviour and beliefs being judged as irrational and illusional. The next chapter considers how the tenets and practices of astrology have coped in a world dominated by cosmos.

CHAPTER 4:

ASTROLOGY, THE QUESTION OF SCIENCE OR SPIRIT

THE OBJECT OF this chapter is to focus on the philosophical world view that gave rise to astrology. It considers the role of astrology in the Old Babylonian culture of the third millennium BCE. It then introduces the reforms to astrology from its Hellenization by Ptolemy, through to the thirteenth-century reductionism of Ramon Lull, and the seventeenth-century reform work of Johannes Kepler. Finally the chapter considers the attempts to scientifically prove astrology in the late twentieth century. This historical overview is undertaken in order to highlight the long and unsuccessful struggle to align astrology with cosmos. The chapter concludes by suggesting that astrology's continual inability to find a home in cosmos may be a consequence of its philosophical absolutism emerging from its chaotic features.

THE PHILOSOPHICAL ORIGINS OF ASTROLOGY

The nature and some definitions of astrology were discussed in chapter one. Here I am concerned with the philosophical world view which shaped astrology. According to Herman Hunger and David Pingree, the earliest acknowledged astrological writings come from the Old Babylonian period of the first half of the second millennium BCE.[109] Hunger and Pingree separated astrology from haruspicy, divination through reading animal entrails, by pointing out that the celestial omens were among those that were unprovoked, independent of a query or human action.[110] James Tester referred to the celestial omens of the Old Babylonian period by commenting that 'They clearly presupposed that there is some relationship between what happens in the sky and what happens on earth, though they do not suggest that the relationship is one of cause and effect'.[111] Christopher Walker argued that with the formation of astrology, the Mesopotamians sought to understand the emerging patterns of events on earth through the mixture of celestial phenomena, weather patterns, and omens.[112] Campion made a similar argument describing

---

109 Hermann Hunger and David Edwin Pingree, *Astral Sciences in Mesopotamia* (Leiden: Brill, 1999), p. 1.

110 Hunger and Pingree, *Astral Sciences in Mesopotamia*, p. 5.

111 James Tester, *A History of Western Astrology* (Woodbridge, UK: The Boydell Press, 1987), p. 13.

112 C. B. F. Walker, 'A Sketch of the Development of Mesopotamian Astrology and Horoscopes', in *History and Astrology*, ed. by Annabella Kitson (London: Uwin Paperbacks, 1989), pp. 7–14, here p.10.

the origins of astrology as a marriage of heaven and earth which developed in a visual format in Mesopotamia where the observable world was read as a flowing, interchanging relationship between the life on earth and the sky.[113]

One of the techniques of this early form of astrology was the use of ritual as an aid to ward off or alter the expression of an ominous celestial omen such as an eclipse of the sun or moon, which represented a serious visual disorder of the heavens. To dispel this evil a bronze kettledrum was played.[114] A letter written in the middle of the eighth century BCE captured this as follows: 'As regards the lunar eclipse about which the king, my lord, wrote to me, it was observed in the cities of Akkad, Barsip, and Nippur. ...A bronze ket[tledrum] was set up...'.[115] Such an action belonged to a philosophy that would now be defined as traditional or religious. Nevertheless in essence it revealed a society that saw itself holding onto order through a continual need for defence against the emergent potential of chaos. This was achieved by using non-linear symbolic or across-scale actions such as beating a bronze kettledrum to remove the chaotic threat of a lunar eclipse.

Another letter written by the priest Akkullānu talked of a difficult celestial omen where an error had been made by the king in the performance of the proscribed ritual. This error was met with anger and frustration by the priest who

113 Nicholas Campion, *An Introduction to the History of Astrology* (London: IS-CWA, 1982), p. 7; Reiner, 'Babylonian celestial divination', p. 22.

114 Hunger and Pingree, *Astral Sciences in Mesopotamia*, p. 6.

115 Simo Parpola, *Letters from Assyrian Scholars to the Kings Esarhaddon and Assurbanipal Part 1* (Germany: Butzon and Kevelaer, 1970), p. 278.

chided the king and his advisors by stating, 'why did they act in this way; did they tell you that your father had placed his statues there? These [talks] are rubbish...Now the king of Akkad [the enemy of the state] is well'.[116] The priest continued by urging the king to perform protective rituals and then apologized for being angry but stressed that, as the king's servant, it was his duty to ward off any threats of incoming evil to the kingdom. These words and actions were not about appeasement or negotiation with a divine force, they are instead, physics. Just as today an industrial national government may adjust its public spending to ward off a potential recession, so the founders of astrology used the beating of a bronze kettledrum, or the erection of substitute statues, to reduce the impact of what for them was the sign of forthcoming evil. One action is based in the logic of Keynesian economic philosophy, the other is based in the 'logic' of chaosmos.

Moreover, celestial omens were just one phenomenon to be observed. Any form of disruption to the order of the world was seen as a meaningful indication of the potential for a new and possibly disruptive pattern to emerge from the chaotic void of Tiamat. Monstrous births, weather patterns, clouds, and the like were all unprovoked styles of omens. Pattern watching for an indication of future events was, and still is for some, a vital tool when living in a world defined by a sympathetic philosophy. With all things bound by a multiplicity of links, then a manifestation across any scale, such as a threatening cloud pattern, could be the first

116 Parpola, *Letters from Assyrian Scholars*, p. 298.

visible emergent sign of a political upheaval in the royal court. Such a style of physics was not exclusive to the Mesopotamians, but to reiterate a point made earlier, this was the philosophical world view of the society that produced astrology. However, this central, non-negotiable pillar of astrology—the need for a sympathetic world view—caused problems for the rationale of astrology once *sumpatheia* was abandoned by later cultures.

COSMOS AND THE ENIGMATIC PRESENCE OF ASTROLOGY

Fuelled by Plato's and Aristotle's thinking, the Greeks took what they saw as the ill-defined, unmanageable Mesopotamian mixture of sky omens and observations and proceeded to develop what is now called horoscopic astrology.[117] For the Greek world, astrology was potentially the 'physics' of the day, the corpus of work that could explain the workings of the world. Thus the old Chaldean astrological thinking needed to be brought into the Greek world view of rational thought. By the end of the last century BCE there was an outpouring of Greek astrological work which, Campion cites, was the foundation for all future branches of astrology.[118] This shift from the Mesopotamian/Chaldean visually based astrology to the Hellenistic horoscopic tradition was one of the earliest pressures placed on astrology to adapt,

---

117 Campion, *An Introduction to the History of Astrology*, pp. 14–16.
118 Campion, *An Introduction to the History of Astrology*, p. 25.

as it was pulled away from its chaotic philosophy of origin and asked to find a home in the order-loving world of the Greeks.

Ptolemy, as one of the most influential astrologer/astronomers in the history of the subject, contributed to this Greek adaption of astrology.[119] Tester claimed that Ptolemy strove to correct and improve the logical order of astrology.[120] Ptolemy's aim was to use Aristotelian principles to build a common-sense foundation for astrology. In his astrological work, the *Tetrabiblos*, he applied logical elegance to astrology and defined the meaning of planets through Aristotelian principles, re-arranging some of the attributes of the zodiac signs to make the system more logical. For example Ptolemy argued that,

> There are two methods of disposing the terms of the planets, in reference to the domination of the Triplicities; one is Egyptian, the other Chaldaic. But the Egyptian method preserves no regular distribution, neither in point of successive order nor in point of quantity.[121]

Terms are a way that a zodiac sign allocated quality and power to a planet (*see* Glossary). Ptolemy, however, disregarded the older 'Egyptian', non-rational doctrine in favour of his more logical system. In this way Ptolemy created an astrology based on a horoscope which owed more to a Pla-

119 Tester, *A History of Western Astrology*, p. 3.
120 Tester, *A History of Western Astrology*, p. 12.
121 Ptolemy, *Tetrabiblos*, Bk. I.33.

tonic perfect or ideal form than to its original free-flowing sky narrative first developed in Mesopotamia. Even though Ptolemy's astrology supported the notion of *sumpatheia* his reforms laid the foundations of the pursuit of a cosmic-rational astrology.

A thousand years later another scientific reforming astrologer, the theologian Ramon Lull (1232–1316), like Ptolemy, also tried to develop a linear rational astrology. He reduced all of the astrological components of a horoscope down to their simplest form, assigned them an alpha-numerical value and then rebuilt them into a form of algebra in order to exact a more consistent or deeper understanding of the astrological patterns.[122] Lull's astrological experiment, although ingenious, was not adopted by later astrologers.

Some few hundred years later Johannes Kepler, discussed earlier, reformed astrological aspecting, the angular separation between planets or horoscopic points (*see* Glossary). Prior to Kepler, the importance of the geometrical distances between planets was not only contained in the physical distance but was also dependent on the planet's 'personality' and 'place of residence'.[123] From a planet's perspective, certain 'places of residence' did not have easy communication with other places. As Masha'allah ibn Athari (c. 740–815 CE), a Persian Jewish astrologer, wrote of an aspect between the Sun and Mars in a particular chart, 'The Sun

---

122 Ramon Lull, *Treatise on Astronomy Books II–V*, ed. by Kristina Shapar, trans. by Robert Hand, (Berkeley Springs, WV: Golden Hind Press, 1994), p. v.

123 Masha'allah, *On Reception* (Reston, VA: ARHAT Publications, 1998), p. 2.

in this aspect does not receive Mars because Mars is not in the Sun's domicile nor in his exaltation, and Mars himself receives the Sun because the Sun is in Mars' domicile'.[124] Thus Mars was willing to be in communication with the Sun but the Sun was not receptive to any possible conversation. This astrological blending of place and aspect was in fact an attempt to reflect the nature of human relationships and their resulting communication networks. Mars, as an individual, seeks to 'talk' or 'work' with the Sun but the Sun is not open to this exchange. The notion of the power of place, the location of a planet within a zodiac sign, is known within astrology as the essential dignities (see Glossary), which Francesca Rochberg-Halton argued have their roots in Babylonian astrological practices.[125]

This Babylonian concept of the essential dignities viewed the planets as members of a community, each with their own property, status, and sphere of influence and, like members of a human community, seeking and owing favours to other planets. Kepler, however, in his pursuit of a mechanical sky, removed all of these relationships and reduced the planets to orbiting objects where motion was the principle concern.[126] With this as his central premise, Kepler devised new aspects which replaced the notion of a planetary community with the logic of geometric configu-

---

124 Masha'allah, On Reception, p. 3.
125 Francesca Rochberg-Halton, 'Elements of the Babylonian Contribution to Hellenistic Astrology', Journal of the American Oriental Society 108 (1988): pp. 51–62, here pp. 53–57.
126 Nicholas Campion, A History of Western Astrology Volume II (London: Continuum Books, 2009), p. 106.

rations. Kepler's reforms had a substantial impact on the practice of astrology, moving it, as Ptolemy had done earlier, still further away from any parallels with the vagaries of life.

By the early twentieth century astrologers, apparently forgetting the central tenet of astrology's need for the recognition of *sumpatheia*, were vigorously pursuing membership to the cosmological world view. Early in the twentieth century the German astrologer Reinhold Ebertin (1901–1988) attempted to return to Lull's reductionism through his development of what he defined as cosmobiology. This system utilised planetary midpoints, the half-way point between planets measured along the ecliptic, and their resulting geometrical 'trees'. In producing these midpoint trees Ebertin removed the meanings of zodiac signs, aspects and houses (*see* Glossary) and stripped astrology down to its bare planetary geometrical relationships. Each midpoint was then given a meaning and thus a 'tree' could be read piece by piece. This was Ebertin's attempt to create scientific astrology and set it apart from what he saw as being its more ancient, imperfect, and vulgar origins. As he said of his new thinking in astrology, 'Cosmobiology is a scientific discipline concerned with the possible correlations between cosmos and organic life and the effects of cosmic rhythms and stellar motions on man'.[127]

A further attempt to apply reductionism to astrology was made in the 1950s by the French couple Michel and François

---

127 Reinhold Ebertin, *The Combination of Stellar Influences* (1940; repr. Tempe, AZ: AFA., 1997), p. 11.

Gauquelin. They were not attempting to reform astrology, but rather they sought to prove astrology within the scientific model, by using tools which needed astrology's cosmic quality-loaded numbers to act as simple integers. Their most famous research result, published in *L'Influence des Astres* (1955), was what became known as the Mars Effect. In this research the Gauquelins claimed a statistically significant finding for Mars to be in a particular arc of the sky at the time of birth of a French sports champion. This rush towards statistics and seeking a place amongst the sciences is also evident in the writings of Doris Chase Doane, who at the same time as the Gauquelins' publications wrote, 'Ideas that go back through ancient Chaldea and ancient Egypt to Atlantis and Mu can be explained in terms of the most modern discoveries of material science'.[128] In order to show that the 'ancient ideas' of astrology could be proved by way of the rigour of modern scientific method, astrological research journals were established to aid the discussion and advancement of what was considered a most important and potentially legitimising work. One of these journals, and probably the most prestigious, was and indeed still is *Correlation*, published twice a year in the UK by the Astrological Association of Great Britain. It first appeared in the autumn of 1968 through to the summer of 1970 when it was co-published with ISAR (International Society for Astrological Research) from Ohio, USA. It then ceased for a period of eleven years, but resumed in June 1981 and has

128 Doris Chase Doane, *Astrology, 30 Years Research* (Los Angeles: The Church of Light, 1956), p. ix.

continued to be published bi-annually since that date. In the first issue editor Simon Best displayed an optimism and willingness to form a partnership with the scientific community, even suggesting that astrology was prepared to change some of its tenets in order to be accepted into the world of twentieth-century science. In 1968 Best commented,

> Science, in its painstaking way, is gradually coming to conclusions that promise to make astrology respectable. However, astrologers, from their intuitive viewpoint, should not be too patronising, their own subject being too imperfect not to need the help that science can give.[129]

What then follows are eight volumes of *Correlation* where only quantitative number-based research projects were undertaken, none of which managed to fulfil the early expectation of that first editorial. By the summer of 1970 Dennis Elwell, the new editor, displayed a shift of thinking and stated, 'If a scientific astrology is to be developed, meticulous attention must be paid to its language. Modern science has been conditioned by linguistic philosophy and the semantic discipline'.[130] Elwell also suggested that astrology may need to change some of its ideas if it was to be accepted by science. Astrology was flexible enough to change any of its ideas except for its absolute requirement: *sumpatheia*. The next publication of *Correlation* was eleven

---

129 Simon Best, 'Editorial', *Correlation* (1968): pp. 1–9, here p. 9.
130 Dennis Elwell, 'Editorial', *Correlation* 8 (1970), p. 3.

years later in June 1981. Newly revitalised by the introduction of computers into astrologers' research work, the journal now took a different approach to the problem and suggested that the failure to find a scientific proof was not the fault of astrology but was instead the fault of astrologers. This was expressed in the words of Professor Hans Eysenck when he wrote,

> For the past few years, Dr David Niss and I have been looking closely at the literature which has been accumulating with respect to the scientific testing of astrological hypotheses. We found much of the material interesting and suggestive, but we are most impressed by the weakness of the scientific methodology employed by most of the authors.[131]

John Addey, however, in the same edition of the new *Correlation*, did not accept that it was the astrologers' fault but rather astrology, as it was practiced, was incompatible with the scientific model. He lamented, 'I have to begin by saying, regretfully, that I am more or less convinced that the astrology of the text-books is, to a considerable extent, in the nature of an elegant fiction'.[132] By this stage the cultural impact of the inability of astrology to produce scientifically acceptable results was being felt by astrologers who were facing either the total rejection of their tenets or the rejec-

---

131 Hans Eysenck, 'The importance of methodology in astrological research', *Correlation* 1 (1981): p. 11.

132 John Addey, 'The True Principles of Astrology and Their Bearing on Astrological Research', *Correlation* 1 (1981): pp. 26–35, here p. 26.

tion of the dream of a scientific astrology. Yet not all astrologers agreed with Addey. Six month later in December 1981, Michael Shallis took up the challenge of defending astrology's boundaries arguing that,

> Modern scientific research and astrology are different kinds of activity. Astrology is derived from given Principles, research from theory and empirical data. The author argues that the two activities come from different world-views and one cannot validly be regarded in the light of the other.[133]

This is an important point made by Shallis as it drew a line in the sand. The community of western astrology, having faced the disappointment of failure in not being proven by the scientific model, was starting to recognise that the problem was deeper than just finding the right database or experiment. Shallis actually raised the point that it was just possible astrology belonged to a 'different world-view'. Addey, unfortunately, died before he could reply to Shallis' challenge.

Correlation continued, however, with yet another ten years of quantitative research into astrology, seeking to gain entry into the scientific world. By this time Michel Gauquelin, with his debatable statistical results for his Mars effect, had become the champion of the astrological community. With the slow corrosion of his research results, however, and his tragic suicide in 1991, a turning

---

133 Michael Shallis, 'Problems of Astrological Research', *Correlation* 1 (1981): p. 41.

point in the attitude of astrologers occurred. The optimism of the community for astrology to be recognised as a science expressed by Doane in the 1950s, was now replaced with anger and/or despair. Patrick Curry, in his memorial to Michel Gauquelin, claimed that astrologers 'were understandably chary of yet another successful scientific operation in which, unfortunately, the patient [astrology] did not survive'.[134] By 1995 this despair turned to resignation with the then editor of *Correlation*, Rudolf Smit, stating:

> The AA set up *Correlation* as a platform of scientific research into astrology, hoping that it would help create an astrology based on fact rather than on assumption. To a large extent this has not come about—despite many published studies, and despite occasional hopeful findings we seem to be no nearer our goal than when we started.[135]

This sterile view that astrology could only be legitimate if it followed the scientific method was also supported by the work of Geoffrey Dean. He, like other astrological researchers in this period, was forgetful of *sumpatheia* and viewed the cosmic number component of astrology as simple integers. Dean thus believed that astrology should conform to the nature of a linear dynamic system, and hence it should be responsive to statistical analysis and reductionism. Dean reviewed astrological research from 1900 to 1976 and discussed and discredited over two hundred separate

---

134 Patrick Curry, 'Memorial to Michel Gauquelin', *Correlation* 11 (1991): p. 9.
135 Rudolf Smit, 'Editorial', *Correlation* 14 (1995): p. 1.

astrological experiments, all of which were quantitative research projects endeavouring to prove astrology by the scientific method.[136]

In the face of such prolonged criticism and failure to gain acceptance by the scientific community, it is understandable that most astrologers would strive to distance themselves and their subject from this reductionist approach. In fact the real result of this eighty years of research into the possibility of astrology being a science is the evidence that it is not. What has been gained, after hundreds of research projects, is the finding that astrology is not a linear dynamic system; it is not a science in the waiting room still anticipating the discovery of its causal agent. Shallis was correct when he stated that astrology belonged to 'another world view'.

The reaction to this result has been for many astrologers to view astrology as a sacred science, a spiritual or religious phenomenon. In this light Garry Phillipson suggested that astrology was a body of knowledge that will not allow itself to be tested, measured, and dissected and that astrologers may need to return to the ancient and traditional idea that astrologers themselves are required to partake in the unfolding of astrology by way of their own purity of heart or holiness, suggesting that astrology may work best when approached as a sacred art.[137] Earlier in 1930, Charles

---

136 See Geoffrey Dean, *Recent Advances in Natal Astrology* (Australia: Fowlers, 1977).

137 Garry Phillipson, 'Astrology and the anatomy of doubt', *The Mountain Astrologer* 104 (2002): pp. 2–12, p. 2.

E.O.Carter had pondered the same problem when, in struggling to understand the nature of astrology, he wrote, 'As within, so without. I know that many astrologers detest what they consider to be mystical; but I see no other rational explanation of facts'.[138] This tension between science and spirit not only exists amongst those who attempt to define astrology, but it has also flowed into the mind-set of its practitioners.

In 2011, Nicholas Campion and Liz Greene commented that 'no two astrologers appear to be able to agree on what it is they believe in, how they define their work, and what metaphysical or religious framework, if any, they espouse to justify what they do'.[139] This is also supported by the research conducted by Phillipson who compiled a cross-section of astrological thought and opinions and provided a window into the views of its practitioners and critics in the late twentieth century. Amongst the different groups of astrologers Phillipson interviewed, there were those who believed that astrology was a science with one astrologer commenting, '...the discovery of the mechanism by which it [astrology] works can only be a matter of time, and likewise it is inevitable that science will eventually recognise that astrology works and welcome it back to the faculty of

---

138 Charles E. O. Carter, *The Astrological Aspects* (1930; repr. Pomeroy, WA: Health Research, 1934), p. 14.

139 Nicholas Campion and Liz Greene, 'Introduction', in *Astrologies: Plurality and Diversity*, ed. by Nicholas Campion and Liz Greene (Ceredigion, Wales: Sophia Centre Press, 2011), pp. 1–15, here p. 13.

respectable studies'.[140] This astrologer represented a group within the astrological community who continued to pursue research using the scientific method and believed that astrology belonged in the cosmological world view of linear order. Other groups of astrologers linked astrology to a form of magic and practised their craft as a form of divination, reflective of Fortune's definition cited earlier. What is apparent is that astrologers are ambivalent about their role, ambivalent about their identity, and ambivalent about their purpose. In the words of Phillipson, 'The picture (of astrology as an art touched by divine anarchy) is not an easy thing to make any kind of sense of, but it is essential to an understanding of how many astrologers think about their subject'.[141]

Curry who offered the neutral definition of astrology cited in chapter one, moved away from his neutrality, considering that belief in astrology, or its practice, was a form of enchantment, or an instrument of enchantment, a way in which humanity encounters mystery, awe, and wonder.[142] His focus was that it was necessary for astrology to be marginalised by science in order for it to maintain this position. Indeed Curry, in the same passage, noted that any success gained by astrology in becoming creditable in the view of science 'would be at the price of its soul'. Here Curry was suggesting that it was not possible for astrology

140 Garry Phillipson, *Astrology in the Year Zero* (London: Flare Publications, 2000), p. 182.
141 Phillipson, *Astrology in the Year Zero*, p. 185.
142 Roy Willis and Patrick Curry, *Astrology, Science and Culture: Pulling Down the Moon* (New York: Berg, 2004), p. 89.

to belong to cosmos, as defined in chapter one, as its 'soul', its central principles, were non-compatible with the linear order required by cosmos. This inability of astrology to comply is important. Astrology, as discussed in chapter one, had always proven itself adaptable to its host culture. Nevertheless its incapacity to make this particular adaptation, despite eighty years of attempts by the astrological community, strongly suggests that astrology is unable to function without the core tenet of *sumpatheia*.

This chapter has focused on the philosophical background of the culture that developed astrology into a global practice. It argued that astrology was the product of a culture embedded in a sympathetic world view and its attempt to domesticate, in some small way, the emergentism of chaosmos. The chapter then discussed astrology's adaptability to the desires of reforming movements considering the work of Ptolemy, the reductionism of Ramon Lull, and the successful de-animating of the planets by Kepler. This adaptability, however, proved fruitless in the attempt in the second half of the twentieth century to justify astrology via the scientific methodology. This failure is notable as in its long history astrology has proven to be culturally adaptable and 'obliging' to requests made of it in the past. Based on this failure, many astrologers concluded that astrology was a spiritual subject, requiring the causal agency of the gods. This chapter concludes, however, that this failure is due to astrology's membership to a secular or non-secular chaotic philosophy. The next chapter looks at the return of chaosmos by considering the findings of chaos theory and complexity science.

CHAPTER 5:

THE RETURN OF CHAOS

THIS CHAPTER GIVES an overview of the twentieth century's re-engagement with chaotic thinking. It also considers what is defined as complexity science and its two-dimensional mathematical form known as a fractal. Throughout the chapter, the components of chaos are related to the life sciences and psychology which view fractals as narratives or mythology, while the strange attractors of chaos are compared with the Jungian concepts of archetypes. The aim of the chapter is threefold: establishing the general concepts of chaos theory and complexity science, comparing it to the vernacular thinking revealed in the chaosmic creation myths, as cited in chapter two, and establishing the nature of chaos in order to allow, in the next chapter, its mapping onto astrology.

In 1899 a French mathematician, Jules Henri Poincaré (1854–1912), solved the Three Bodies problem first posed by Newton.[143] Newton had been seeking the solution to predicting the future position of three planets, each of which applied gravity to each other. Newton had found the limits of classical science, defined by Capra as causal and linear.[144] Poincaré solved Newton's problem by mathematically proving that there was no solution, that the positions could not be predicted, only estimated.[145] Poincaré's solution, however, went largely unnoticed as physics, at that time, was enthralled with the work of Albert Einstein (1879–1955), who was shortly to become a deified figure of science with his Theory of Relativity.[146] Nevertheless Poincaré is now considered to be the grandfather of chaos theory, as his solution of 'no solution' was the beginning of the undoing of Newton's clockwork universe. Poincaré had shown that science could not *know* everything. The paradigm of the ordered cosmos had encountered its limitations. Laplace's argument was finally, albeit slowly, being banished. As Bütz commented, the undoing of the mechanical world of order was not caused by some obscure cerebral-rarefied problem. It was instead undone through the seemingly simple situation of understanding the orbital path of three planets all influencing each other by gravity.[147] Thus was

---

143 Isaac Newton, *The Principia* (Berkeley, CA: University of California Press, 1966), Prop. 66, Bk. 1.

144 Capra, *The Web of Life*, p. xxviii.

145 See Henri Jules Poincaré, *Science and Method* (London: T- Nelson, 1914).

146 Bütz, *Chaos and Complexity*, p. 6.

147 Bütz, *Chaos and Complexity*, p. 6.

the weakness in the linear model exposed. According to Mainzer, this was a turning point from ordered cosmos back to non-linear chaos: 'The equilibrium of the Parmenidean world broke down and changed to the evolutionary and complex world of Heraclitus'.[148] Poincaré's 'no solution' solution gave the mental jolt which opened the door just wide enough for chaosmos to be glimpsed.

LINEAR AND NON-LINEAR SYSTEMS

At this point the distinction between linear and non-linear systems needs to be clarified. Simply put, a chain of dominoes is a linear dynamic system. A tap on the first domino will cause it to fall forward and strike the second domino and consequently the second domino will fall onto and strike the third domino, and so on, along the chain until finally the last domino will fall. Science totally understands this event. Taking into account the size, weight, and distance between each domino, the fall of the last domino is not only assured, but also its time of falling can be accurately predicted. Such a system is defined as linear, as it is activated one step at a time while being driven by clear antecedent causes—in this instance the tap to the first domino which then 'runs' through the chain. In contrast to this is a non-linear dynamic system. A crowd of people in a limited space is an example of such a system. If the crowd begins to surge, the final end position for any one

148 Mainzer, *Thinking in Complexity*, p. 71.

individual cannot be predicted because people are jostling one another as they move. Such a system is in feedback and this feedback can be used to calm and order the crowd (each person resisting pushing into others) or it can be used to stampede the crowd, (each person applying force to others in order to create room for themselves). Whatever the nature of this feedback, the end result for a given individual or for the entire crowd cannot be predicted using classical science. This is defined as a non-linear dynamic system and, although this is a simple example it can be viewed within the domain of chaos theory.

A BRIEF OVERVIEW OF CHAOS THEORY AND COMPLEXITY

The term chaos was first used in science by the American James Yorke in 1975. He used it in its classical scientific expression referring to the findings of Edward Lorenz a researcher from MIT.[149] Lorenz had found that in weather prediction small changes in variables led to vastly different results, a finding that defied prediction in classical science as it was a non-linear response. Yorke's use of the term chaos described the unpredictable and apparent randomness in behaviour of complex interlinked systems and he used it to express all the negative cultural values that chaos and non-order had come to mean to the western world.

Briefly, as stated by Prigogine and Stengers, chaos theo-

149 Gleick, *Chaos: Making a New Science*, pp. 65–69.

ry is the study of the impact of feedback into a non-linear dynamic system.[150] Such systems are considered chaotic, random, and unable to be studied by the reductionist methodology of classical science. Capra addressed the problem of understanding non-linear dynamic systems by stating:

> This new approach to science [complexity and chaos] immediately raises an important question. If everything is connected to everything else, how can we ever hope to understand anything? Since all natural phenomena are ultimately interconnected, in order to explain any one of them we need to understand all the others, which is obviously impossible.[151]

Here Capra, or the science that he represented, stood in the same position as Plato when he was dealing with the implications of Heraclitus' philosophy discussed in chapter three. The dilemma presented by Heraclitus was that if all things are in a state of change, if all things are linked, then one could never really know anything.[152] Capra, however, proposed the solution of 'approximate knowledge':

> The old paradigm is based on the Cartesian belief in the certainty of scientific knowledge. In the new paradigm it is recognized that all scientific concepts and theories are

---

150 Prigogine and Stengers, *Order Out of Chaos*, p. 75.
151 Capra, *The Web of Life*, p. 40.
152 Heraclitus, *Heraclitus Fragments*, fg. 41.

limited and approximate. Science can never provide any complete and definitive understanding.[153]

Capra's idea contained two important points. The first was that since 'approximate knowledge' accepted the connectedness of all things, the inter-relationships and feedback of one system on another, it allowed space for the principles of chaos and complexity to be acknowledged as valid concepts. Furthermore, it challenged the value judgement—'perfect', 'imperfect'—that science had placed on knowledge. Capra's solution effectively did away with Plato's and later René Descartes' split of the world of two separate places, that of perfect and that of imperfect.[154] Thus at least in principle there was no longer a need for conflict between the forces of cosmos and chaosmos. This potential removal of the tension between order and chaos was vital, for research indicated that the chaotic non-linear dynamic systems were the very systems which described all of life. Whether 'life' was defined as a simple, single cell or a complex animal or even a society of organisms functioning in their environment, it was self-evident that life functioned in relationships.[155] R.A.Thiétart and B. Forgues echoed this when they claimed that this new concept (chaos) discovered by science was proving to be the creative source for all living systems, as it is the source of

153 Capra, *The Web of Life*, p. 40.

154 René Descartes, *Principles of Philosophy* (1644; repr. New York: Springer, 1984), Pt. IV, art. 187.

155 R. A. Thiétart and B. Forgues, 'Chaos Theory and Organization', *Organization Science* 6 (1995): pp. 19–31, here p. 19.

self-organising of new patterns.[156]

The acceptance of chaos theory was not a smooth process. In 1989 Robert Poole, seeking to dismiss chaos theory or belittle its findings, turned to the earlier classical definition of chaos as disorder and defined it thus: 'Chaos is order disguised as disorder, a sheep in wolf's clothing'.[157] Poole's arguments come from classical science's success in understanding linear dynamic systems and its assumptions that these solutions could be applied to all problems. For the findings based in linear dynamic systems had traditionally been applied to problems involving living systems. For example, in talking about the field of economics, Mitchell Waldrop pointed out that the goal in economies was that of stability or 'balance'.[158] Similarly John van Eenwyk pointed out that this desire for order and stability was the preferred state for organisations, political systems, populations, communities, and even in individual lives.[159] However, research into chaos, the behaviour of non-linear dynamic systems, found the exact opposite, that stability could in itself lead to stagnation.

156 Thiétart and Forgues, 'Chaos Theory and Organization', p. 19.

157 Robert Poole, 'Is It Chaos, or Is It Just Noise?', *Science* 243 (1989): pp. 25–28.

158 Mitchell Waldrop, *Complexity: The Emerging Science at the Edge of Order and Chaos* (New York: Touchstone, 1992), p. 17.

159 John van Eenwyk, *Archetypes and Strange Attractors: The Chaotic World of Symbols* (Toronto: Inner City Books, 1997), p. 43.

TWO TYPES OF CHAOS — ENTROPIC
AND DETERMINISTIC

By the late twentieth century economists discovered that
stable systems tended towards stagnation or death or what
is now called entropic chaos.[160] An example of a simple
form of entropic chaos is a pendulum clock with its pendulum swinging back and forth over a fixed point which
eventually winds down. Its energy becomes the same as
that of its environment and movement no longer occurs.
As Paul Davies noted, however, if systems are pushed into
an unstable situation through feedback, then new patterns,
new order, and new options were encountered.[161] This was
called deterministic chaos and has been applied across
many diverse disciplines from economics and companies
to family therapy and even to the health and lifestyle of individuals.[162] Experience shows, however, that it is unwise
to push a stable, orderly but becoming static economic or
social system into complete chaos as the human suffering
involved in the process of the new emerging economic or
social patterns is usually considerable. When this occurs
in political systems it is called a revolution and, although
this eventually yields new patterns, it does so with loss of

---

160 See Waldrop, *Complexity*.
161 P. Davies, *The New Physics* (New York: Cambridge University Press, 1989), p.
501.
162 Rae Fortunato Blackerby, *Applications of Chaos Theory to Psychological
Models* (Austin, TX: Performance Strategies Publications, 1993), pp. 135–37;
Thiétart and Forgues, 'Chaos Theory and Organization', p. 19; Eenwyk, *Archetypes and Strange Attractors*, p. 45; Bütz, *Chaos and Complexity*, p. 18.

life and general human suffering. Thus in the 1990s economists were studying just how much an economy needed to
be 'disturbed' in order for it to start to yield new patterns.
Tim Clark, in discussing Whiteheadian philosophy spoke
of the idea of a 'domesticated chaos', which he defined as
having the value of 'loosening restrictions, avoiding the
ponderous, and combat rigidity'.[163] This domestication of
chaos is more aptly described as complexity, discovered as
the state of liminality which exists in the place between the
abyss of chaos and the order of cosmos, and now seen as the
place of creation.

COMPLEXITY — A STATE OF BECOMING

There is a thin zone which exists between a static system
and a system in chaos. It is in this thin zone, defined by
Kauffman as the 'edge of chaos' where complexity increased
and order naturally occurred.[164] This discovery of a natural
increase of order, seen as new patterns forming in the zone
between static and chaotic, occurred within systems which
were in feedback, such as the crowd of people discussed
earlier. Such systems were in contrast to those which did
not contain feedback, such as the line of dominos, as these
systems 'wound down', becoming simpler rather than com-

---

163 Clark, 'A Whiteheadian Chaosmos', pp. 101–2.
164 Stuart Kauffman, 'Antichaos and Adaptation', *Scientific American* 256 (1991):
       pp. 78–84; Kauffman, *At Home in the Universe*, p. 15; also see E. Corcoran,
       'The Edge of Chaos', *Scientific American* 267.A (1992); Waldrop, *Complexity*.

plex. This phenomenon produced by feedback, was called *Complexity*.[165]

The application of complexity and its implications for living systems has caused it to spread beyond the field of economics. Life, like an economy, engages in feedback, creates webs of naturally-occurring interlinking patterns within its environment and forms social and/or community systems such as hives, anthills, nests, colonies, flocks, herds, tribes, cities, and nations.[166] An example of complexity thinking is to consider the earth's biosphere as a place of 'complexity'. It is a thin film wrapped around the earth about twenty miles thick. In proportion to the earth, the biosphere is no thicker than a coat of paint on a globe the size of a soccer ball. It is also the place where all life on earth exists. Thus all life on this planet exists in this thin space between the solidness of the earth and the vacuum of space. And according to complexity, it is in this zone we live our lives in an order of repeating patterns.[167] Similarly the way an individual organism moves through its life—being influenced by and influencing its environment—means that it, too, will be involved in these naturally occurring patterns and experience them as upheavals and disturbances, times of ease and times of stress.

Research in complexity and chaos thinking has also been absorbed into understanding human social systems,

---

165 Paul Cilliers, *Complexity and Postmodernism: Understanding Complex Systems* (1998; repr. London: Routledge, 2005), pp. 2–3.

166 Kauffman, *At Home in the Universe*, p. 186; Abraham, *Chaos, Gaia, Eros*, p. 209.

167 Kauffman, *At Home in the Universe*, p.15; Capra, *The Web of Life*, pp. 209–11.

such as family therapy and business systems.[168] Mary Ann McClure commented that:

> Self-organizing government agencies function best in conditions of bounded instability, instead of a state of equilibrium, as advocated by the more traditional system theorists. It follows that managers must accept uncertainty and risk and learn to see crises as opportunities that allow for innovation and a new responsiveness. They must learn not only to be comfortable with instability, but also to create disorder when the organization becomes too stable.[169]

It is now accepted that chaos, through complexity, can create new order and in turn order is not stable and can move towards chaos. The place of creation is at the 'edge of chaos' before the system disintegrates, as this is where the greatest complexity will occur, in the form of new patterns, new order. These new patterns continue to grow and manifest until eventually they, too, will move back into chaos and once again re-emerge transformed into new order and patterns (deterministic chaos), or end, never to appear again (entropic chaos). This behaviour is indifferent to scale and can be seen in the single life of an individual or in the his-

---

168 Bütz, *Chaos and Complexity*, p. 18; John D. Eigenauer, 'The Humanities and Chaos Theory: A Response to Steenburg's "Chaos at the marriage of heaven and hell"', *The Harvard Theological Review* 86 (1993): pp. 455–69, here p. 469.

169 Mary Ann McClure, 'Chaos and feminism—A Complex Dynamic: Parallels Between Feminist Philosophy of Science and Chaos Theory', (2004), http://www.pamij.com/feminism.html [Accessed 5 October 2004].

tory of a nation. Such a nation can emerge out of times of upheaval and chaos, take root and grow and then change or eventually disappear, remaining only in history. In this way the new chaos of mathematics is reflective of the mythic, primal, and creative chaosmos of the old, disregarded creation mythology.

FRACTALS AND THE PATTERNS OF LIFE

The role that chaos and complexity play in understanding the emerging dynamics of an individual life or a whole living system can be understood using the visual form of complexity—fractals.

Benoit Mandelbrot (1924-2010) coined the term 'fractals' (*see* Glossary) from the Latin *fractua*—to be irregular.[170] He was working with the hand-calculated and hand-drawn graphs of the early twentieth century French mathematician Gaston Julia (1893-1978). The patterns/graphs produced by Julia and published in 1918 were a visual representation of how order was created at the edge of chaos. Julia found these complex patterns when he iterated an equation; iteration is the repeated feedback into an equation of its solution. Julia worked with imaginary numbers (*see* Glossary) which are those that retain their quality of either negative or positive values independent of the calcu-

---

170 Benoit Mandelbrot, *Fractals: Form, Chance, and Dimensions* (San Francisco: W.H. Freeman, 1977).

lations in which they are used.[171] Imaginary numbers are akin, in philosophical nature, to the cosmic numbers contained in astrology as they also consist of two parts, the planetary quality (meaning) and movement (numerical value). Julia found that when such numbers were subject to feedback then patterns emerged. The actual patterns formed by Julia were created by plotting the results of the iterations of the equation onto a graph as a dot. The resulting image is a fractal produced by an iterated equation and is a simple visual representation of a non-linear dynamic system as it moves through time (see Fig. 5).

FIGURE 5. A Julia fractal formed by iterating an equation and plotting the results each time on a graph, after thousands of iterations a pattern begins to appear. Image generated by the fractal program 'Ratio'.

171 Edward S. Allen, 'Discussions: Definitions of Imaginary and Complex Numbers', *The American Mathematical Monthly* 29 (1922): pp. 301–3, here p. 302.

Julia, and later Mandelbrot, found that by observing the imagery of fractals, the shapes exhibited scale invariance and grew in a self-similar manner.[172] To clarify this point, a fractal will continue to produce the same patterns and images regardless of the scale or magnification (Fig. 6). When the results of the equation eventually move off into infinity (entropic chaos) this is represented by the black or grey area in the diagram in Figure 6, left. The 'edge of chaos' is the area where the visual pattern occurs. When the pattern was magnified (Fig. 6, right) it revealed the repeating theme of the set, regardless of the scale of magnification.

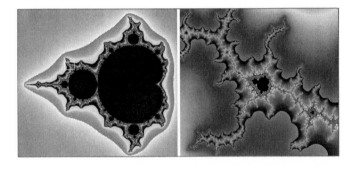

FIGURE 6. A fractal known as the Mandelbrot set. The image on the left is the whole set, while the image on the right is a magnification of the set, which shows that, after extreme magnification, the shape of the original set reappears in the pattern. Image created by author using Fractal Forge.

172 Bütz, *Chaos and Complexity*, p. 17.

Mandelbrot recognised two points regarding work in fractals and their relationship to an understanding of living systems. The first was that the shapes of nature could be reproduced by iterating these types of equations. These are now called fractal forgeries where the images of a leaf, tree, mountain range and so forth can be formed by the emerging fractal (Fig. 7).[173] Hence it was recognised that the building blocks of nature, the shapes made, consisted of sets of functions which could be represented by genetics and/or the environment, which were in a continual feedback loop. The second point was that living and non-living forms in nature also grow through what was defined as self-similarity and scale invariance.[174] The fern leaf was made up of a lot of smaller fern leaves all repeated, all a little different from each other but all reproduced in a self-similar manner. Additionally it was found that shapes were reused by nature for different objects, thus a river system was similar in shape to that of the bronchi of a lung or the shape of a cauliflower stem. Nature was filled with fractals across both living and non-living units regardless of scale. The different shapes in fractal mathematics could be produced by using different equations. Tiny changes in the variables used at the beginning of the iterations of the equation gave vastly different results in the end pattern. This sensitivity to small changes in the initial variables be-

173 Image from Fractal Forgeries: http://classes.yale.edu/fractals/IntroToFrac/
    SelfSimEx/SelfSimEx.html [Accessed 5 October 2004].

174 Eenwyk, *Archetypes and Strange Attractors*, p. 54.

came known in chaos theory as Sensitive Dependence on Initial Conditions or SDIC.[175]

FIGURE 7. An example of an equation which, when fed back on itself, produced points on a graph that 'drew' a black spleenwort fern. *Wikimedia Commons.*

The step that Mandelbrot took with Julia's work was to produce a formula which accounted for all possible imaginary numbers that, when iterated, did not instantly move into infinity. Mandelbrot found how to produce all possible fractals, all possible shapes of nature, in one set. This product or image is now known as the Mandelbrot set (Fig. 6) and is a collection of all possible patterns. It is Mother Nature's master template in which patterns and shapes in the world, as we know it, reside.[176]

The implications of fractals and thus complexity's

175 Eigenauer, 'The Humanities and Chaos Theory', p. 458.
176 William Keepin, 'Astrology and the New Physics', *The Mountain Astrologer* (Aug/Sept., 1995): pp. 12–21.

findings of self-similarity and scale invariance have been drawn into psychology. Fractals are simple visual maps which show how complex systems in feedback will produce new order. Yet this system could be a company with many employees or the life of a single individual, as both are systems which are sensitive to feedback. In discussing the use of chaos and complexity in the practice of psychology Michael Bütz wrote:

> Self-similarity adds understanding to this type of change process, which moves from stable to bifurcation [a point of change] to chaos or complexity and into a new, more complex and adaptive order. These changes become a universal process as systems become unstable. This process can be applied to an individual, a family, a community, and so on—in a self-similar manner.[177]

He is not alone with these ideas. The effect of complexity and chaos on the field of psychology has also been described by the Jungian psychologist van Eenwyk who commented, 'There is a recognition that reductionism cannot be used to understand the human psyche, that there is no longer an assumption that order is healthier than chaos. There is a paradigm shift occurring in the psychological landscape in terms of chaos'.[178] This paradigm shift referred to by van Eenwyk is a new way of understanding how life moves through time.

177 Bütz, *Chaos and Complexity*, p. 18.
178 Eenwyk, *Archetypes and Strange Attractors*, p. 43.

Fractals show, in a graphic manner, that life 'lives' by repetition or daily rituals (iteration) of personal behaviour patterns which produce events of similar nature (self-similar) across the many areas of a person's, company's, or nation's life (scale invariance). In psychology this has led to the now-recognised phenomenon of families reproducing the same style of events over different generations.[179] Until Mandelbrot's work in fractals, these self-similar events were disregarded as little, idiosyncratic, non-explainable, shoulder-shrugging patterns: the third son killed in war for three generations, the members of a family all born or dying on the same day of the week, daughters mirroring mothers in giving birth to the same number of children, or divorcing at the same age as their mother, or in the number of marriages they made, the gender of their children, and so on. As Margaret Ward commented, a normal family will function in patterns because it is a living unit which is subject to feedback.[180]

Both Abraham and Bütz argued that our group and individual lives emerge in the same manner as fractals, and indeed these family or personal reproductions of patterns or events, far from being seen as odd, meaningless and at

179 Murray A. Straus and Richard Gelles, *Physical Violence in American Families: Risk Factors and Adaptations to Violence in 8,145 Families* (New Brunswick: Transaction Publishers, 1990), p. 245; also see Susan Forward and Craig Buck, *Toxic Parents: Overcoming the Legacy of Parental Abuse* (London: Bantam, 1990).

180 Margaret Ward, 'Butterflies and Bifurcations: Can Chaos Theory Contribute to Our Understanding of Family Systems?', *Journal of Marriage and the Family* 57 (August, 1995): pp. 629–38, here pp. 630–31.

times considered superstitions, are actually to be expected and should be considered the norm.[181] Van Eenwyk extended this thinking into the life of a single human being. He argued that the repeating images of a fractal can be recognised as being equivalent to the personal habits of the individual. These may be stuck states in the person's life where they run the same patterns: similar relationships with similar problems, or always finding they are lucky or unlucky. Van Eenwyk makes an argument for the scale invariance of a fractal being the individual's tendency to apply their philosophy of life to their political ideas, preferred sports, hobbies or personal neuroses or even to the small patterns of how one brushes one's teeth.[182] Scale invariance can also be expressed in how an individual exists as part of a family, community, or nation—as an event appears in these communities, so the individual responds.[183] Thus an event on one scale is linked to an event on a smaller or larger scale.

BIFURCATIONS—THE WAY THAT CHANGE ENTERS LIFE

A fractal also has many different branches, twists and turns which can explode into new forms of the inherent pattern or spin off out of the pattern and disappear into in-

181 Abraham, *Chaos, Gaia, Eros*, p. 215; Bütz, *Chaos and Complexity*, p. 4.
182 Eenwyk, *Archetypes and Strange Attractors*, pp. 129–37.
183 Helene Shulman, *Living on the Edge of Chaos: Complex Systems in Culture and Psyche* (Zurich: Daimon, 1997), pp. 13–14.

finity—the black or grey space outside the image as shown in Figure 6. Such a twist or turn is called a bifurcation (*see* Glossary), which is a junction where the equation can jump in either direction, and it is not possible to predict the direction.[184] There are different types of bifurcations which can produce different parts of the pattern. One particular bifurcation of note is called a hopf bifurcation named after its discoverer, the Polish mathematician Heinz Hopf (1894–1971), and it is a bifurcation which can potentially push the pattern in an entirely new expression.[185] A hopf bifurcation occurs at what is now called a saddle point (*see* Glossary). At a saddle point two pathways are offered to the ongoing iterated equation, one of which will lead to an end, the other path leading to new patterns, new images, and new order. Both possibilities are an expression of chaos. When the equation 'chooses' the pathway to infinity, it leaps to a place in which it disappears (into the black area of Fig. 6) and, as previously stated, the equation is said to move into entropic chaos. However, when the equation takes the path to new patterns, this is an example of deterministic chaos, the pathway of new possibilities.

These findings are also applied to psychology. A person encounters choices or events throughout their life. If, however, they seek too much order or stability, they will avoid or ignore the hopf bifurcations (options for change). This

---

184 Ward, 'Butterflies and Bifurcations', p. 631.

185 See Steven H. Strogatz, *Nonlinear Dynamics and Chaos: With Applications in Physics, Biology, Chemistry, and Engineering* (Reading, MA: Addison-Wesley, 1994).

action of shunning hopf bifurcations (options for change) tends to result in life heading into entropic chaos (stagnation). The person's life becomes one where no new opportunities present themselves, no new ideas and no new patterns are allowed to enter their world, and the person's life becomes non-changing all the way to eventual death. In contrast to this, if a person seeks or accepts hopf bifurcations by actively engaging in the periods of disorder where new patterns can emerge (coincidences that lead to new opportunities), then the person's life will expand with new incoming ventures and new ideas.[186] At times the new patterns encountered by the individual will be older patterns on a smaller or larger scale, while other patterns will take the individual to a new part of their life, a new part of their 'personal fractal'. This would be seen as a new area of life but one that was always inherently there within the initial set of forming conditions (SDIC).

RITUALS, SUPERSTITION AND OMENS

The practice of rituals has also been linked to the nature of fractals and thus to chaos and complexity science. Van Eenwyck suggested that the practise of a ritual could be reduced to the simple idea of the iteration of an equation.[187]

---

186 See Bütz, *Chaos and Complexity*; Mainzer, *Thinking in Complexity*; Eenwyk, *Archetypes and Strange Attractors*. For this phenomena in organisations see Ralph D. Stacey, *Complex Responsive Processes in Organizations: Learning and Knowledge Creation* (London: Routledge, 2001), pp. 174–76.

187 Eenwyk, *Archetypes and Strange Attractors*, p. 113.

He argued that by considering that time is circular not linear, then we can use ritual at seasonal times to remake the world and allow humankind to rejuvenate itself, just as chaos rejuvenates order through repetition, as the repeating nature of ritual is the self-similar nature of life. Additionally an individual can use the concept of self-similarity in a more mundane manner. If a particular bifurcation (event) yields pleasing results, then the person could consciously set up the pattern and run it again and again. Colin Campbell argues that this repetition gives rise to habits, rituals, and superstitions at the individual or collective level of society.[188] He suggested that people will consciously reproduce a pattern which has previously proved fruitful and although classical science labels this type of action as superstitious, as in the example of Waugh's red handkerchief in chapter three, chaos and complexity, however, actually suggest that it may be a valid methodology.

ATTRACTORS — THE GHOST IN THE MACHINE

The actual emerging patterns produced by a given fractal are generated by what chaos theorists call a strange attractor (*see* Glossary).[189] There are three main types of attractors. The simplest is a point attractor: a funnel with all the

---

188 Campbell, 'Half-belief and the Paradox of Ritual Instrumental Activism', p. 154.

189 Douglas Polley, 'Turbulence in Organizations: New Metaphors for Organizational Research', *Organization Science* 8 (1997): pp. 445–57, here p. 446.

fluids running to the centre is a point attractor. A periodic attractor is one where bodies have a periodic oscillation like the planets orbiting around the sun, where the sun is the periodic attractor. The third attractor is called a strange attractor. Capra points out that strange attractors exist in a chaotic system and can be understood by observing the external patterns of an iterated equation or life (person, ant, company, economy, and so on).[190] The emerging events of the equation or life will appear to occur randomly in a manner which looks, to the person embedded in classical science, as being unpredictable. Nevertheless, when this journey of emergence is plotted graphically, the movement, far from being random, actually orbits around a set of multi-dimensional foci. The moving foci are called strange attractors. Lorenz's famous butterfly and weather example, cited earlier, came about when he plotted the apparent random results of his iterated equations and produced an image in the shape of a butterfly. Each 'eye' on a wing was the focus point of a strange attractor (see Fig. 8).[191]

These strange attractors are more than just the reasons for an emerging pattern. Complexity theorists such as van Eenwyk claimed that the interlinking of life with all its environmental influences actually created strange attractors and these in turn produced fractal patterns, patterns of behaviour.[192] Indeed as C. Middleton, G. Fireman and R. Di-

---

190 Capra, *The Web of Life*, p. 132.
191 Ian Stewart, *What Shape is a Snowflake?* (New York: The Ivy Press Limited, 2001), p. 177.
192 Eenwyk, *Archetypes and Strange Attractors*, p. 48.

Bello argued, strange attractors were an invisible hand in the events of life, directing these events in what appeared to be random but were instead falling into a precise set of patterns.[193] Van Eenwyk actually saw the concept of Jungian archetypes as the result of the influence of strange attractors in an individual's or community's life. He suggested that narrative, myths, fairy stories, and folklore could be seen as verbal fractals which showed different choices (bifurcations), some leading to loss and destruction (entropic chaos), others leading to fruitfulness (deterministic chaos). The end result of such a story was that often a new strange attractor was found and taken back to the tribe.[194] Bütz also addresses this point and identified the influence of strange attractors in a person's life as issues that could pull them into different life attitudes. He wrote, 'It seems in psychology or clinical psychology more exactly, that an attractor can be roughly equivalent to a healthy or an unhealthy agenda that pulls an individual in one way or another.'[195]

In the same way that Bütz referred to strange attractors influencing an individual, they could also influence a family and be experienced as a family curse or a family gift. A historical example of this was the seventeenth-century German musical family, the Bachs. Over two hundred years the family produced over fifty notable musicians and com-

---

193 C. Middleton, G. Fireman, and R. DiBello, 'Personality Traits as Strange Attractors'. Paper presented at, *Inaugural Meeting for the Society for Chaos Theory in Psychology* (1991), San Fransicso, CA.

194 Eenwyk, *Archetypes and Strange Attractors*, p. 120.

195 Bütz, *Chaos and Complexity*, p. 21.

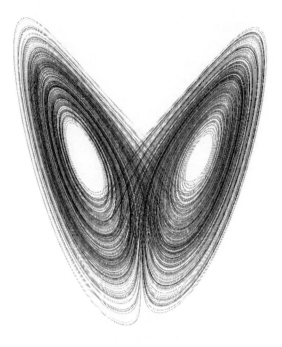

FIGURE 8. The Lorenz strange attractor with its two 'butterfly eyes'. Image from Ian Stewart article 'The Lorenz Attractor Exists', Nature Journal 31 (August 2000): p. 949.

posers.[196] One can argue that the Bach family contained a powerful musical strange attractor. In contrast, a family that was continually beset with illness or general misfortune over several generations could, in this model, be said

196 See entry for 'Bach' in *The New Encyclopaedia Britannica*, ed. by Philip W. Goetz (Chicago: Encyclopaedia Britannica, Inc.), pp. 539–44, here p. 539.

to have within their midst a more difficult strange attractor.

To the non-psychological individual such an 'invisible hand', to use Middleton, Fireman and DiBello's description, begins to take on what Sarah Broadie described as the experience of fate in our lives. Broadie argued that fate feels like the presence of '...an agency concerning which all we know is that it has a certain purpose...and is clever and foreknowing and powerful enough to carry out this intention no matter what'.[197] Robert Solomon also referred to this idea of purposefulness with fate when he suggested that it has an 'otherness' to it, writing that 'Fate provides something by way of "hands" in which we can place ourselves whether or not we have confidence in their goodness'.[198] Fate can have many expressions and this is not the place for a consideration of its history. Suffice it to say, however, elsewhere I have shown that fate, at least within the views of contemporary astrologers, is thought of as an agency that forms self-similar patterns.[199] This suggests that one view of fate is that it is the lived experience of strange attractors in a person's, family's or even nation's life.

Returning to the ideas of myth and fairy story, the fact that they can be discussed in the language of complexity and chaos—the hero following a hopf bifurcation into de-

197 Sarah Broadie, 'From Necessity to Fate: A Fallacy?', *The Journal of Ethics* 5 (2001): pp. 21–37, here p. 28.

198 Robert Solomon, 'On Fate and Fatalism', *Philosophy East and West* 53 (2003): pp. 435–54, here p. 442; For a similar argument also see Sophie Botros, 'Freedom, Causality, Fatalism and Early Stoic Philosophy', *Phronesis* 30 (1985): pp. 274–304, here p. 279.

199 Brady, 'Theories of Fate Among Present-day Astrologers', p. 198.

terministic chaos resulting in a strange attractor which is returned to the village as a gift—does not change the fact that the myth or fairy story will still contain a valuable moral lesson. This lesson can be learned by the listener and the verbal fractal (story) used to help move or prepare the listener for when they personally encounter that particular hopf bifurcation (a life event that requires a particular choice) in order that they take the pathway that leads to the enrichment of their community's life (deterministic chaos). As van Eenwyk argued, nothing about life is changed by this new language of chaos and complexity; rather science is now finding a language for the ways of life.[200]

This chapter has discussed the major components of chaos and complexity science and their extension of fractals. Additionally it has shown how the findings in these areas have been absorbed into the life sciences and psychology. The next chapter extends this mapping onto the practices and beliefs of astrology.

200 Eenwyk, *Archetypes and Strange Attractors*, p. 120.

CHAPTER 6:

MAPPING ASTROLOGY TO CHAOS AND COMPLEXITY

THE AIM OF this chapter is to map the findings from the fields of chaos theory and complexity science onto contemporary astrology. This is undertaken to show that astrology could be considered a vernacular attempt by humanity to bring a level of domestication to chaos, a way of living with and in chaosmos. The chapter first considers the components of self-similarity and scale invariance in astrology, and then argues that the horoscope reflects what is now defined in chaos theory as a phase portrait, a map that gives information about a qualitative non-linear dynamic system. The chapter concludes with a summary in table format of the major parallels between astrology and chaos, and thus by extension astrology and chaosmos.

QUALITY RATHER THAN EXACTNESS: LIFE
AS COMPLICATED

In an interview with Phillipson, English astrologer Michael
Harding stated, 'I think astrology is infinitely complex, I
think life is infinitely complex. If astrology claims to mirror life, then it will have life's complexity'.[201] Here Harding
was using the word 'complexity' in its non-mathematical
sense, although without intention, he may have been offering an insight into the nature of astrology.

To reiterate, astrology works on the principle of scale
invariance fuelled by the 'physics' of *sumpatheia*. Thus
the patterns of the heavens are related to the patterns of
a person's life, one example of this scale invariance being
the seventh-century BCE Venus omen cited earlier which
talked of a green Venus spelling death for unborn children.[202] Such scale invariance—sky events linked to an
individual life—has been central to astrology from its in-
ception to the present day.[203] Added to scale invariance is
the idea of self-similarity. Research into complexity shows
that in non-linear dynamic systems a pattern may be repeated but exactness cannot be predicted. Capra talked of
the predictability of chaos as, 'It is impossible to predict the
values of variables of a chaotic system at a particular time,
but we can predict the qualitative features of the system's

---

201 Phillipson, *Astrology in the Year Zero*, p. 187.

202 Reiner, 'Babylonian Celestial Divination', p. 33.

203 For modern examples see Dane Rudhyar, *The Lunation Cycle* (Santa Fe, NM:
     Aurora Press, 1967), p. 1; Alan Leo, *The Progressed Horoscope* (1905; repr.
     London: L.N. Fowler, 1929), p. 17.

behaviour'.[204] Middleton, Fireman, and DiBello, referring to the phenomenon of complexity, reflected a similar idea, 'Exact behaviour would be unpredictable from moment to moment but it would remain within loose boundaries— those of the strange attractor'.[205] Many astrologers also believe that the horoscope can be used to define the quality of the *pattern* formed by the planets, rather than the exact *event* or *characteristic*—a self-similarity style of thinking where things are repeated but not exactly as they were before. Dane Rudhyar (1895-1985) captured this when he discussed astrological predictive work: 'Events cannot be foretold accurately, but the conditions needed by an individual if he is to grow to his full stature as an individual can be pre-diagnosed. The astrologer can discover from the progressions the main turning points in the life of a person'.[206] Rudhyar's comments were not in isolation within astrological literature.[207] Indeed in my own research into fate and astrology I found that, in surveying over a thousand astrologers, 89% of female astrologers and 83% of male astrologers considered that astrology could not predict an exact event, only its quality.[208] Thus for both the complexity scientist and the predictive astrologer, although using

204 Capra, *The Web of Life*, p. 137.

205 Middleton, Fireman, and DiBello, 'Personality Traits as Strange Attractors', p. 19.

206 Rudhyar, *The Lunation Cycle*, p. 104.

207 See for example Liz Greene, *The Art of Stealing Fire* (London: CPA Press, 1996), p. 147; Darrelyn Gunzburg, *Life After Grief: An Astrological Guide to Dealing with Loss* (Bournemouth, UK: Wessex Astrologer, 2004), p. 82.

208 Brady, 'Theories of Fate Among Present-day Astrologers', p. 187.

very different tools, both view prediction as only the 'loose boundaries' of an emerging pattern.

SELF-SIMILARITY AND SCALE INVARIANCE— THE ASTROLOGICAL USE OF CYCLES

The concepts of self-similarity and scale invariance are also reflected in how astrologers use cycles. Within astrology, cycles are based on planetary periods and then linked, by scale invariance, to the life of the person (a country, an organisation, and so on). Along with other disciplines, astrologers view all cycles as expressions of the same rhythm: a beginning, a period of growth, a time of fullness, and then a period of waning leading to a 'winter' or an end point which in turn leads to a rebirth. Whether the cycle is the diurnal rhythm of a day and night over twenty four hours, or the monthly rhythm of the lunar phases, or the cycles of outer planets—returns or aspect—or the synodic cycles of two planets which may take hundreds of years to repeat their relationship, all cycles in astrological lore move through the same rhythm.[209] Thus within the principles of self-similarity and scale invariance, astrologers will observe an event which occurs at one point of the cycle as re-emerging in essence at the same point in a later expression of the cycle.

209 Rudhyar, *The Lunation Cycle*, p. 3.

PHASE PORTRAITS AND HOROSCOPES

Additionally, along with the themes of order manifesting through the lens of scale invariance and self-similarity both astrology and complexity also accept that the future is inherently contained within the system. For astrologers, the 'system' is the map of the heaven for the time of birth, while complexity links it to the more general overview of the nature of the initial conditions. The astrologer's map, the horoscope, is a tool which plots the image of the solar system for the moment and place of an event, birth or otherwise. This map is the astrologer's 'ordered' version of the 'initial conditions'. Central to astrology is the belief that the patterns of sky and earth at the beginning of the individual's life can provide information about the emerging patterns for the rest of that individual's life. This is based on scale invariance and what was defined in the classical period as *sumpatheia* which has been discussed in detail earlier. Dane Rudhyar, reflecting the opinion of many astrologers, talked of the horoscope as representing a complex structure linked with life that was not subject to reductionism and thus could not be broken into parts.[210]

This idea of a life map which is based on the initial conditions is also one that is central to chaos theory and complexity as it attempts to understand the emerging nature of a non-linear dynamic system. These maps are called phase portraits (*see* Glossary) and were defined by Capra as a map of the emerging potential of a particular pattern which can

---

210 Rudhyar, *The Lunation Cycle*, p. 5.

be used to gain understanding of the quality of a non-linear dynamic system. He stated, 'The qualitative analysis of a dynamic system, then, consists of identifying the system's attractors and basins of attraction, and classifying them in terms of their topological characteristics. The result is a dynamical picture of the entire system, called the "phase portrait"'.[211] Capra, talking purely of mathematics, was stating that an emerging pattern could be mapped in terms of its attractors. Psychology would call these attractors the personality, make-up, family issues, and parts of the self that pull events and people into the life.

One of the attributes of an attractor of any type is that they have basins, an area of influence. For a funnel (a point attractor) the basin is the width of its mouth. In terms of an individual life it will also contain attractors whose basins will be the person's influence in their work, family, and the lives of others. A person's life, like any other non-linear dynamic system, can contain many attractors and thus even potentially overlapping or competing basins. The 'topological characteristics' referred to by Capra in his definition of a phase portrait measure how flexible and adaptable the system is to its attractors. In psychology this is how flexible the individual is to what the attractors, via their basins, bring into the person's life. The maps, or phase portraits, are used for the qualitative analysis of the 'system'.[212] In the same way one can argue that the horoscope is a vernacular expression of a phase portrait of a person's life. It contains

211 Capra, *The Web of Life*, p. 134.
212 Capra, *The Web of Life*, p. 134.

planetary patterns which, according to astrologers, act as attractors and, as suggested by the work of Rudhyar, cited earlier, astrologers attempt to read the quality of the 'system' from these maps. The following are examples of the way in which the astrological features of a horoscope echo the mathematical features of a phase portrait.

PLANETARY PATTERNS AND STRANGE ATTRACTORS

All systems subject to chaos create or contain strange attractors, and horoscopes contain what astrologers call planetary patterns. These planetary patterns are the geometrical shapes formed between the planets and the surface of the earth at the time of a person's birth. Depending on the type of astrology being practised, these are thought by the astrologer to be particular markers or tendencies within the person's life or within the person's psyche which will dictate/symbolise/indicate their behaviour patterns. Astrologers will also assign orbs to a planetary pattern which can be classed as one way to measure the basin of an attractor. Indeed before aspects were modernised by Johannes Kepler, the orbs used by astrologers were orbs applied to each individual planet. Planets acted like spheres on a net, some 'heavier' than others but all created indents and therefore inter-relating basins of attraction in the horoscope.[213] In this way the planetary positions in the horoscope can be reflective of the strange attractors that chaos and complexity indicate occupy living systems.

HOPF BIFURCATIONS —
THE RANGE OF EXPRESSION WITHIN A PLANETARY COMBINATION

Many astrologers working with a horoscope will discuss with a client the productive uses of a feature in the horoscope versus its non-productive or self-negating use. They will look at an upcoming potential and suggest ways that the client can 'make the most' of the opportunity and even talk of negative options that the client may wish to avoid.[214] This is similar to what complexity would define as the potential of a hopf bifurcation (opportunity for change) to develop into its deterministic (creative outcome) or entropic (stagnation outcome) chaotic expressions. Complexity also indicates that at a hopf bifurcation one may be able to predict the quality, however the actual pathway taken — entropic or deterministic chaos — cannot be foreseen. This concept was expressed by mathematicians Maturana and Varela in their discussion on predicting such outcomes at bifurcation points: 'When a living system reaches a bifurca-

213 Guido Bonatti, *Liber Astronomiae Part II* (Berkeley Springs, WV: Golden Hind Press, 1994), p. 4.

214 See for example Michael R. Meyer, *A Handbook for the Humanistic Astrologer* (New York: Anchor Press, 1974), p. 21; Gunzburg, *Life After Grief*, p. 136.

tion point, its history of structural coupling will determine the new pathways that become available, but which pathway the system will take remains unpredictable'.[215] This is the mechanism behind the concept of self-similarity; the choice made by the 'living system' is not predictable but its 'structural coupling' will define the options available to the life. Astrologers would agree that seeing the possible options open to a person of an indicated forthcoming period contained a theme, defined by the natal horoscope and the quality of the planetary meaning being mathematically applied to that horoscope, but would not be able to determine the exact action the person will take.

SADDLE POINTS AND HOMEOSTASIS—
THE SENSITIVITY OF HOROSCOPIC POINTS

Astrologers will find possible future events by using mathematical ways of moving the horoscope through time and locating the particular planetary pattern which is being influenced at that time. Astrologers consider that each point in the horoscope is sensitive, in its own way, to bringing change into a person's life. In a phase portrait, chaos labels such points as saddle points, the place where the pattern can change its shape. Additionally the sensitivity of the system is known within chaos and complexity thinking as its level of equilibrium. Living systems display what is

---

215 Humberto Maturana and Francisco Varela, *The Tree of Knowledge* (Boston: Shambhala, 1987), p. 95.

known as homeostasis, the ability to maintain an equilibrium and be resistant to small disturbances (*see* Glossary).[216] This potential for stability and change within a homeostatic system can be measured in chaos theory by the size of the strange attractors and the nature of the saddle points. With this in mind chaoticians are creating a bifurcation encyclopaedia.[217] In non-chaos language, this can be expressed as the size of the area of influence that the system draws on (basin), the nature of this influence, and the sensitivity of the system to this influence. In a human life this resistance to change is not solely at the biological level, for a person also has boundaries which are material, emotional, spiritual, and intellectual. Astrology reflects these concepts of saddle points and homeostasis, for astrologers consider that some parts of a horoscope are more sensitive to astrological predictive events than others, and some sorts of predictive events are more powerful than others. Astrologers, like chaoticians, also provide a corpus focused on defining the horoscopic sensitivities and which many published works on predictive astrology reflect.[218]

Thus the horoscope can be viewed as a vernacular attempt to create a phase portrait of a complex system in order to judge what Capra argued was 'the qualitative analysis of a dynamic system'. I make no claim to astrology's efficacy

---

216 Kauffman, *At Home in the Universe*, p. 79.

217 Abraham, *Chaos, Gaia, Eros*, p. 61.

218 See for example Robert Hand, *Planets in Transit* (Rockport, MA: Para Research., 1976); Reinhold Ebertin, *Directions: Co-Determinants of Fate*, trans. by Linda Kratzsch, (Tempe, AZ: AFA, 1976); Nancy Anne Hastings, *Secondary Progressions: Time to Remember* (York Beach, ME: Samuel Weiser, 1984).

CHAOS AND COMPLEXITY: MATHEMATICAL FINDINGS	COMMON PRACTICES AND BELIEFS OF ASTROLOGERS
Sensitive Dependence on Initial Conditions (SDIC). *The initial conditions dictate the emerging patterns.*	*The moment of birth will influence the emerging pattern of a person's life. The horoscope.*
Phase Portraits: an image of the attractors, basins and flexibility of a pattern which can be used to predict the times and qualities of changes to the pattern.	*The creation of a horoscope which can be used to observe the emerging patterns and timing of these patterns in a person's life.*
Strange Attractors—the moving foci which seem to invisibly influence the external events or pattern.	*Planetary combinations which define the nature or quality of what the person will attract towards themselves and the story of their life.*
Hopf Bifurcation—a change in the pattern which can lead to new patterns or entropic chaos—death.	*A time in a person's life indicated by a predictive event where the individual can be encouraged to take the path which yields the most creative options.*
Saddle Points—the point where hopf bifurcations occur.	*Sensitive points in the horoscope which, when receiving some form of predictive event, will result in events of a certain quality occurring in the person's life.*
Self-similarly and scale invariance. Repeating themes in patterns occurring in unrelated patterns—e.g. river systems and the bronchi of lungs.	*Astrologers use of cycles which link planetary cycles with smaller cycles within an individual (or country or organisation) life.*

TABLE 1. Comparison between the features of chaos and complexity and the practice of astrology.

in this endeavour. I do, however, argue that astrologers, in attempting to create an image of a living system, created the horoscope which contains the same *components* as a phase portrait—planets used like attractors, orbs used like attractor basins, horoscope points seen as saddle points. Additionally the way that astrologers use this horoscope is reflective of the use of a contemporary use phase portrait—stability, predictability of pattern rather than exact event, and points of varying sensitivity. How these attributes were interpreted at different periods of astrology's history have vary greatly however, the horoscope and the use of planets and aspects are enduring features of astrology.

In conclusion, the points made in this chapter are summarised in Table 1, which shows some of the possible links between the mathematics of chaos and complexity and the practice of astrology.

All of the above correspondences could be challenged in different ways. One could contest the very findings of chaos and complexity, or one could argue that the vernacular maps of astrology are too crude and simplistic to have any relationship to the maps of chaos. Such criticism, however, would be missing the point, for what is central here is not the veracity of astrology but rather the argument that over its history astrologers have intuitively developed an approach to seeking order in life. Astrologers have developed tools which seem to allow them to sit on the 'edge of chaos' and stare into the emerging patterns of chaosmos. My hypothesis is that, independent of arguments of usefulness, these tools bear a strong resemblance, in a vernacular manner, to those developed in chaos theory and complexity science.

Additionally, although one can show that humans think, live, and create in the manner of complexity which is subject to SDIC (Sensitive Dependence on Initial Conditions), it is only astrologers who claim that the heavenly patterns into which a human is born constitute part of those initial conditions. There can never be a statistical proof of this point, for as Capra states about the new science of non-linear dynamics, it is made up of 'Qualitative instead of quantitative, Relationships instead of objects, and Patterns instead of substance'.[219]

The next chapter brings my research to a conclusion with a discussion of the major arguments and implications.

219 Capra, *The Web of Life*, p. 113.

CHAPTER 7:

ASTROLOGY AND CHAOSMOS

CONCLUSION

THE AIM OF this project has been to explore the possible links between the practice of astrology and the views emerging from the findings of chaos theory and complexity science. In this endeavour the work followed other academic pursuits which have examined chaosmic elements in the fields of literature, philosophy, and music by investigating the works of certain individuals, notably James Joyce, William James, Alfred Whitehead, Gilles Deleuze, and Umberto Eco.[220] I, however, have focused on the subject of astrology, rather than the work of a single individual. If astrology can

---

220 See for example, Farronato, *Eco's Chaosmos*; Clark, 'A Whiteheadian Chaosmos'; Ruf, *The Creation of Chaos*; Bidima, 'Music and the Socio-historical Real'.

be located in chaosmos, then such an alignment potentially allows for new insights concerning the nature of astrology and the role it plays in contemporary society.

In reviewing the arguments, the first chapter established the parameters of chaosmos and cosmos and the attributes of astrology. It argued that there were two central tenets to astrology, the first being that of cosmic numbers, numbers consisting of both quality and quantity, while the second was the notion of *sumpatheia*, the physics by which these complex numbers gained the efficacy to 'speak' of life on earth. This work then considered creation mythologies and showed how these myths described the creation of order and, in doing so, also established the definition of rational thought for their respective cultures. Order from chaotic creation myths was revealed as emergent, polymorphic, and non-anthropological in orientation and based in cycles, while what was considered to be 'rational' in these myths included omens and superstitions. These myths were in contrast to the cosmic creation myths with their linear modes of generating order which produced a definition of the 'rational' as that which was linear and predictable. Then by exploring the historical journey of the privileging of the cosmic style of creation, I argued that this led to the ontological properties of chaosmos being forgotten or dismissed as illusional and irrational.

Chapter four examined the creation mythology of the people of Old Babylon who developed astrology as an important tool for maintaining order in the kingdom. These people belonged to a culture embedded in a sympathetic world seeking to domesticate, in some way, the emergentism of chaosmos. Thus astrology's philosophical rationale was founded within a chaotic world view. This link of astrology with the *sumpatheia* of chaotic creation myths was/is central to astrology. Thus even though astrology, in part, is culturally relative as it has shown to be adaptable to the desires of reforming movements—from the work of Ptolemy, the reductionism of Ramon Lull, and the successful de-animating of the planets by Kepler. This relativism disappeared, however, when its chaosmic elements were challenged by the attempts to have it accepted into the rational world of ordered cosmos.

Nevertheless although astrology could not conform to the requirement of ordered cosmos, its hybrid numbers, loaded with planetary symbolism, were confused with normal integers, and by forgetting the requirement of *sumpatheia* these numbers suggested astrology could confirm to cosmic order. This, when linked with the apparent blind acceptance of the dominance of cosmos, has led to ambiguity around the nature of astrology. Such confusion is evident in the observation by Campion and Greene, cited earlier, that 'no two astrologers appear to be able to agree on what it is they believe in, how they define their work'.[221] With this ambiguity established, chapter five considered the return of chaos as an ontological force in contemporary thinking. Primarily using the findings of the work in fractals, I discussed how these findings had moved from mathematics into the life sciences and psychology. Chapter six then moved from the life sciences to astrology and showed

221 Campion and Greene, *Sky and Symbol*, p. 13.

the parallels between the findings within chaos and complexity and the contemporary practices within astrology.

Underneath all of these arguments lies the notion that several thousand years ago human intuition was capable of conceiving the main principles of chaos and complexity science. Evidence for this, I believe, lies in the chaotic creation mythology discussed in chapter two and the classical principles of *sumpatheia* defined by the Stoic philosophers of the third century BCE discussed in chapter one. For example, the intuitive knowledge of chaos is visible in the Egyptian creation myth of Khnum, the polymorphic water deity who was called Father of Fathers and Mother of Mothers. The attributes of Khnum are those of random emergence from a void, which once it begins to take form, then stimulates other potential patterns and coaxes these into a higher form of order (life). Khnum the potter, I contend, sat on 'edge of chaos' like the membrane described by complexity science. This figure periodically returned to the void only to remerge again, not as identical but as what the Egyptians called 'a return to the first occasion'. This deity was an artisan-god but could also be a human artisan (ordered being) who was engaged with chaos, harvesting the opportunities that spontaneously emerged. This is the artist looking for creative inspiration, or the single individual putting their energy into a new emerging possibility. In this way Khnum was/is the personification of many of the attributes which have only recently been ascribed to life by the findings in chaos and complexity—cyclic in his/her return to the first occasion, and engaged with the spontaneous emergence of patterns. Furthermore, in turning to the notion of *sumpatheia*, the chaotic elements of scale invariance and self-similarity are a requirement for the working of omens, the logic of superstitions, and the efficacy of divination.

In recent times authors like Eco and Joyce have intentionally turned to chaosmos as a literary device to inject complexity and the texture of life into their prose. In contrast, the philosopher Alfred Whitehead rather than seeking chaosmos endeavoured instead to build a philosophy based on 'becoming' rather than 'being'. His philosophy, now defined as process philosophy, is used as the philosophical foundation of chaos and complexity.[222] Following these examples, this work suggests that astrology was not *intentionally* designed as a chaosmic discipline but rather emerged as an attempt by human instinct to craft a tool to help domesticate the apparent vagaries and co-incidences of life, a tool designed for gazing into the potent void of chaosmos.

Additionally, my research question on astrology and its relationship to chaosmos could only be asked once the all-encompassing perception of cosmology had been revisited. For this reason I returned to Plato's view of cosmology described in the *Timaeus*, which was a world formed by Intellect and Reason and all linked by the World Soul. This web of influence of the World Soul was hierarchical and intellectual for Plato privileged god and reason above all else and hence supported the creation of order through

---

222 David Ray Griffin, 'Process Philosophy of Religion', *International Journal for Philosophy of Religion* 50 (2001): pp. 131–51, here p. 131.

an artisan, a pre-existent entity with a conscious aim. This is Khnum without the Nile, for Plato's god had no need for contact with the chaotic void, and thus rendered the creative void of chaosmos at least impotent and at worst non-existent. As previously argued this was quite dissimilar to the later Stoic notion of *sumpatheia* which did not privilege any concept, god, human, or nature.[223] It is only by returning to this Platonist view of cosmos that an argument could be made for the existence of chaosmos. Only then could chaosmos' distinctive ontological features be addressed. Indeed by extension it is apparent that in the contemporary period cosmos has become increasingly privileged, possibly by the combined drives of humanism, pluralism and logically, post-modernism. Thus in seeking chaosmos I have argued for the curbing of the boundaries of this new all-consuming version of cosmology. Because such a cosmology removes the potential for the existence of other ontologies, while at the same time it permanently locates the patterns experienced in a lived life to the realms of the non-rational or irrational, belonging to superstitions, illusions or 'old wives tales'.

Chaosmos, however, is now visible. As an ontological force it has been examined in literature, considered in music, and acknowledged in strains of different philosophical thought. Its features are the creation of meaning through increasing and maintaining a high level of connectivity in a system which is sensitive to feedback. When the system is the world as a whole, I argue that this gives rise to what the

223 Laëtius, *Physics*, p. 62. Cicero, text 2.19.

Stoics defined as *sumpatheia*. Its procedure for the creation of new order is cyclic, yet it does not produce a return of the same but rather similar events or symbols which will be reproduced across a wide pallet, independent of scale or form. Viewed from within or from without, chaosmos will appear, as noted in chapter one, to produce a plethora of ideas, coincidences, and events which can fracture across a myriad of reflective 'surfaces' as signs, omens, and coincidences, all of which 'deepen' and enrich the experience of a lived life.

This chaosmos is, I argue, the ontology within astrology and has been so from the Babylonian period to the current day. Therefore it is not sufficient to define astrology purely by reference to cultural relativism, for even though it has adapted to cultural pressures over its long history, it has always resisted abandonment of its chaotic credentials. In truth astrology can embrace the shifting whims of fashion and culture because its focus is not on the culture that life creates but rather the experience of being alive. Curry talked of the 'soul' of astrology being lost if it conformed to science; I agree and suggest that the astrological soul is one which is held in the hands of chaosmos. Chaosmos because the foundations of astrology are chaotic, rational numbers which can be logically manipulated yet are loaded with planetary qualities and given agency through the notion of *sumpatheia*. By acknowledging these essential components, this work argues astrology can provide, no matter how roughly, a logical meaning to the stream of co-incidences, superstitions, happenstances, and synchronicities that constitute a life lived in or near chaosmos.

Glossary of Technical Terms

Attractors: There are three types of attractors:

a) **A point attractor:** This is a system that moves towards a stable equilibrium such as a clock pendulum.
b) **A Periodic attractor:** This is a system with a periodic oscillation. The planets orbiting around the sun move around the attractor but do not encounter it.
c) **A Strange Attractor:** This is a chaotic system. The behaviour of a particle or object appears to move chaotically. However when, it is analysed, the object is found to be moving around a 'moving' group of foci and the movement or the different positions of the foci create a pattern.[224]

See 'Phase Portraits' for an overview of the system of attractors, basins, saddle points and SDIC.

Basin: The area of influence of an attractor. With a funnel,

---

224 Capra, *The Web of Life*, p. 132.

which is a point attractor with its centre hole, the basin of the attractor is the funnel's bowl. See 'Phase Portraits' for an overview of the system of attractors, basins, saddle points, and SDIC.

Bifurcations and Hopf Bifurcations: Points in a dynamic scheme where changes occur to the behaviour of the scheme.[225] A Hopf bifurcation occurs when a bifurcation fails to stabilise itself, leading to a cascade of bifurcations.[226] In psychology a bifurcation is a small event which the individual absorbs into their life causing little change. A hopf bifurcation is a small or large event that causes the whole life to change.

Chaos: Deterministic Chaos and Entropic Chaos: Chaos, as used in chaos theory, refers to deterministic chaos, which is a form of chaos within which patterns periodically appear.[227] In contrast, entropic chaos is when all order is lost and no patterns occur. A burning campfire reducing all to ash is a form of entropic chaos. However, if the burning fire randomly left other shapes and forms after it had died out, then that would be an example of deterministic chaos.

Complexity: A phase membrane which exists between the states of static (no change) and chaos (all change). In this phase membrane spontaneous order emerges which in-

---

225 Abraham, *Chaos, Gaia, Eros*, p. 66.
226 Eenwyk, *Archetypes and Strange Attractors*, p. 61.
227 Eenwyk, *Archetypes and Strange Attractors*, p. 45.

creases the complexity in the system.[228] Complexity is considered a feature of chaotic systems. Complexity science is the study of the behaviour of these complex, self-organising systems in living and non-living forms.[229]

Complex Numbers: see Imaginary numbers.

Differential Equations: Equations which express certain formulas of constant relationships and in which changes in the value or magnitude assigned to certain variable factors determine the value or magnitude of the other variable factors. These equations are helpful in solving many problems of higher mathematics and the natural sciences because the knowledge of certain known factors permits one to compute the value or magnitude of the unknown variable factors.[230]

Entropy: A term introduced by Rudolf Clausius, a German physicist and mathematician, to measure the dissipation of energy into heat and friction.[231] The greater the entropy of a system, the greater the distribution of energy within that system.

Fractals: In 1977 Benoit Mandelbrot coined the term 'fractal'

228 See Corcoran, 'The Edge of Chaos'; Kauffman, *At Home in the Universe*; Waldrop, *Complexity*.
229 Shulman, *Living on the Edge of Chaos*, pp. 16–17.
230 Differential Equations https://www.mises.org/easier/D.asp [Accessed 5 October 2004].
231 Capra, *The Web of Life*, p. 180.

to define one of mathematics 'empirical study of nature'. A fractal is a graph created by plotting the results of equations that are subject to iteration and as a result of this continual iteration are pushed into chaos. The results of an iterated equation will either move off into infinity or produce order (a closed shape). The individual results of the order-producing equations are plotted to produce a visual representation which, as already stated, Mandelbrot defined as a fractal.[232] In this way the visually observed patterns of the fractals are the patterns which are spontaneously formed at the edge of chaos. The concept of fractals has moved beyond mathematics—an example of their apparent universal nature was put forward by John Briggs and F. David Peat who showed that some poetry could and did use this notion of iteration.[233]

Imaginary Numbers: These are numbers which exist outside our normal use of numbers. For example, the square root of 25 is 5 or -5 because the quality of +/- is not maintained in the calculation. However, if the question is, what is the square root of a negative number like -25, the solution cannot be found. To resolve this limitation on the conventional number system the idea of an imaginary number was created. Using an imaginary number the square root of -25 becomes 5i or -5i. Though these numbers do not exist in the realm of real numbers, these imaginary numbers play a vital role in physical and engineering calculations, as well

232 Mandelbrot, *Fractals: Form, Chance, and Dimensions*, p. 2.
233 Briggs and Peat, *The Turbulent Mirror*, pp. 195–96.

as in the generation of Julia and Mandelbrot set fractals.

Imaginary numbers can be combined with real numbers to form what is called complex numbers. The range of complex numbers which do not move into infinity can be plotted on a Cartesian plane (graph) to produce a fractal. The complete set of all of these complex numbers that produce a closed pattern is used to create the Mandelbrot set.

Iteration: Iter is Latin for a journey. The process of feeding the results of an equation back into itself and then re-solving the equation. When some particular forms of equations are 'iterated', it gives rise to chaos.[234]

Homeostasis: A property of a unit to be resistant to small perturbations. Attractors provide a homeostatic situation. In a large system the attractor drains a large basin and so changing any one component within this large basin has little impact within the network. However a unit with a small trajectory and a small basin can be altered permanently by a small change. Any system which does not have homeostasis, is considered to be chaotic.

Phase Portrait: The map which plots or identifies the system's attractors and basins of attraction and classifies them in terms of their topological characteristics. The result is a dynamic picture of the entire system, called the 'phase portrait'.[235]

---

234 Briggs and Peat, *The Turbulent Mirror*, p. 57.
235 Capra, *The Web of Life*, p. 134.

A simple allegory of a phase portrait is the layout of a golf course with its 18 holes as the attractors to which all the golf balls are drawn. The greens and fairways would be the basins—the area that seems to draw in the golf balls towards the attractors. The teeing off area could be considered the saddle points—the place where change will occur. The sand traps and roughs could be the areas of entropic chaos where the golf balls never reappear. The entire golf course plan represents the entire phase portrait of the course. This of course ignores weather conditions and the skill of the golfer, which could be equated to the SDIC factor of the chaotic the system.

Saddle Point: The point in the system where a hopf bifurcation can occur. See 'Phase Portraits' for an overview of the system of attractors, basins, saddle points and SDIC.

Scale Invariance: The feature of a fractal to produce the same shapes regardless of scale. This is also the feature of nature to produce the same shapes from the micro to the macro. In the human sciences it is the feature of a family's history to be similar to that of a country, the events concerning the family pet reflecting the events happening to the family's finances and so forth. Self-similarity and scale invariance are, in chaosmos, the potential mechanism contained within omens, superstition, and divination of all kinds.

Self-similarity: The feature of a fractal to continue to produce similar shapes. In the human sciences it is the feature of a family or a person to experience the re-occurrence of past patterns.[236] Self-similarity linked to scale invariance can be considered within chaosmos thinking, as the logic contained within omens, superstition, and divination of all kinds.

Sensitive Dependence on Initial Conditions (SDIC): This is abbreviated to SDIC. Small changes at the beginning can lead to greatly different results—as in the butterfly effect, which was one of the early discoveries about chaos.[237] See 'Phase Portraits' for an overview of the system of attractors, basins, saddle points, and SDIC.

Structural Coupling: The past history of the equation in the way that it has moved through different bifurcations. In the human sciences this can be equated to the history of the subject.

Topology: Topology is the mathematical study of the properties that are preserved through deformations, twisting, and stretching of objects. It studies the 'wholeness' properties of an object rather than its parts. One of the central ideas in topology is that spatial objects like circles and spheres can be treated as objects in their own right, and knowledge of objects is independent of how they are 'represented' or 'embedded' in space. For example, the statement 'if you remove a point from a circle, you get a line segment' applies

---

236 Bütz, *Chaos and Complexity*, pp. 17–18.
237 Robert W. Batterman, 'Defining chaos', *Philosophy of Science* 60 (1993): pp. 43–66, here p. 48.

just as well to the circle as to an ellipse and even to tangled or knotted circles, since the statement involves only topological properties.[238]

ASTROLOGICAL TERMS

Aspects: The geometric relationship between planets, points, or luminaries. The types of geometric relationships and the orbs used in these relationships vary considerably over the unfolding history of astrology. In the second century CE Ptolemy used just five aspects. Later, Johannes Kepler (1571-1630) introduced new thinking into the field of aspecting and not only increased the number of aspects but disregarded the sign boundaries.[239]

Essential Dignities: A system of allocating a type of affinity between a planet or luminary and a sign. There were five essential dignities recognised in Hellenistic astrology which were in order of strength: rulership, exaltation, triplicity, term, and face. These appeared to be duplicated in Islamic astrology of Abū Rayḥān al-Bīrūnī (973-1048), as well as they passed unchanged into the period of Guido

---

238 http://mathworld.wolfram.com/Topology.html [Accessed 9 September 2004].

239 Fred Gettings, *Dictionary of Astrology* (London: Routledge & Kegan Paul, 1985), p. 173.

Bonatti in the thirteenth century, although Bonatti also called them fortitudes.[240] When a planet was in a zodiacal position in which it is in one of its essential dignities, it was considered to have dignity.

Houses: The method of dividing the horoscope into a total of twelve sections called houses. Each house refers to a different part of a person's (or company, nation, or so forth) life, and each house will be governed by a series of astrological considerations.

Planetary Patterns: A geometrical arrangement of planets in a horoscope which astrologers believe blend, in some fashion, the symbolic meanings of the planets.

Planetary Midpoints: The geometric midpoint between two planets. First discussed by Ptolemy in the second century ce but used extensively by Reinhold Ebertin as the foundation of his reform of astrology, which he called cosmobiology.[241] Midpoints can also give rise to midpoint trees, which are simply the collection of planetary midpoints occurring at the same degree of the zodiac or aspecting it.

Predictive Systems: Any method of moving the horoscope through time. The most common used in modern astrology is probably that of transits: the current position of a planet in its orbit is traced on the horoscope and when it

240 Bonatti, *Liber Astronomiae Part II*, p. 10.
241 See Ebertin, *The Combination of Stellar Influences*.

reaches a sensitive zodiac degree in the person's horoscope, the astrologer gauges that some form of event will occur.[242] The quality of the event will depend on the nature of the horoscopic point, the nature of the transiting planet, and the geometry by which the transit occurs.

Synodic Cycles: Synodic cycles in astrology are the time periods required for two orbiting planets to create their same geometric relationship to each other. The most commonly known synodic cycle is that of the sun and the moon which produces a new moon every 29.5 days. Other synodic cycles are much longer, with that between Neptune and Pluto being 492 years.

Terms and Term Rulers: Terms are seemingly irregular divisions of the degrees of each zodiac sign into groups. Each grouping within a sign is linked to a different planet known as the Term Ruler. Each zodiac sign contains five terms with each one being ruled by one of the five planets. The luminaries do not rule any terms. There are two main types of terms: the earlier set was labelled by Ptolemy as Egyptian and the later set, which were devised by Ptolemy in his attempt to make them more logical, are known as the Ptolemaic terms.[243]

Transits: see Predictive Systems.

---

242 See Hand, *Planets in Transit.*
243 Ptolemy, *Tetrabiblos*, p. 47.

Triplicities: Each sign of the zodiac is assigned one of four possible elements (air, fire, earth, and water) so that there are three signs for each element. Each group of three signs are then assigned three planets which are called the triplicity rulers. For example, for the fire signs the triplicity rulers were: the Sun, Jupiter, and Saturn. However, in an effort to reform what he believed was an illogical assigning of rulers Claudius Ptolemy in his Tetrabiblos allocated only two triplicity rulers per sign instead of the older concept of three.[244]

244 Ptolemy, *Tetrabiblos*, p. 44.

Bibliography

Abraham, R. *Chaos, Gaia, Eros*. New York: HarperSanFrancisco, 1994.

Addey, John, 'The True Principles of Astrology and their Bearing on Astrological Research'. *Correlation* 1 (1981): pp. 26–35.

Al-Biruni. *The Book of Instruction in the Elements of the Art of Astrology*. London: British Museum, 1934.

Allen, Edward S. 'Discussions: Definitions of Imaginary and Complex Numbers'. *The American Mathematical Monthly* 29 (1922): pp. 301–3.

The Astrological Journal. Edited by Zach Matthews and Suzanne Lilley-Harvey. London: The Astrological Association of Great Britian, 1988.

Batterman, Robert W. 'Defining Chaos'. *Philosophy of Science* 60 (1993): pp. 43–66.

Best, Simon. 'Editorial'. *Correlation* (1968): pp. 1–9.

Bidima, Jean-Godefroy. 'Music and the Socio-historical Real', pp. 176–95. In *Deleuze and Music*. Edited by Ian Buchanan and Marcel Swiboda. Edinburgh: Edinburgh University Press, 2004.

Blackerby, Rae Fortunato, *Applications of Chaos Theory to Psychological Models*. Austin, TX: Performance Strategies Publications 1993.

Bobzien, Susanne, *Determinism and Freedom in Stoic Philosophy*. Oxford: Clarendon Press, 1998.

Bonatti, Guido. *Liber Astronomiae Part II*. Berkeley Springs, WV: Golden Hind Press, 1994.

Botros, Sophie. 'Freedom, Causality, Fatalism and Early Stoic Philosophy'. *Phronesis* 30 (1985): pp. 274–304.

Brady, Bernadette. 'Theories of Fate Among Present-day Astrologers'.

PhD Dissertation, University of Wales Trinity Saint David, 2012.

Brady, Bernadette. *The Eagle and the Lark—A Textbook of Predictive Astrology* (Portland, ME: Samuel Weiser, 1992.

Briggs, J., and F. D. Peat. *The Turbulent Mirror: An Illustrated Guide to Chaos Theory and the Science of Wholeness*. New York: Harper & Row, 1989.

Broadie, Sarah. 'From Necessity to Fate: A Fallacy?'. *The Journal of Ethics* 5 (2001): pp. 21–37.

Bütz, Michael. *Chaos and Complexity—Implications for Psychological Theory and Practice*. Washington, DC: Taylor and Francis, 1997.

Campbell, Colin. 'Half-belief and the Paradox of Ritual Instrumental Activism: A Theory of Modern Superstition'. *The British Journal of Sociology* 47 (1996): pp. 151–66.

Campion, Nicholas. 'Astrology's Place in Historical Periodisation', pp. 217–54. In *Astrologies, Plurality and Diversity*. Edited by Nicholas Campion and Liz Greene. Ceredigion, Wales: Sophia Centre Press, 2011.

———. *A History of Western Astrology Volume II*. London: Continuum Books, 2009.

———. 'Introduction', pp. 1–3. In *Cosmologies*. Edited by Nicholas Campion. Ceredigion, Wales: Sophia Centre Press, 2009.

———, *An Introduction to the History of Astrology*. London: ISCWA, 1982.

Campion, Nicholas, and Liz Greene. 'Introduction', pp. 1–15. In *Astrologies, Plurality and Diversity*. Edited by Nicholas Campion and Liz Greene. Ceredigion, Wales: Sophia Centre Press, 2011.

Capra, Fritjof. *The Web of Life: A New Scientific Understanding of Living Systems*. New York: Doubleday, 1996.

Carter, Charles E. O. *The Astrological Aspects*. 1930. Reprint, Pomeroy, WA: Health Research, 1934.

Casey, Edward S. *The Fate of Place: A Philosophical History*. Berkeley, CA: University of California Press, 1998.

Chamberlin Roy B., and Herman Feldman, eds. *The Dartmouth Bible*. Boston, MA: Houghton Mifflin, 1950.

Chapple, Christopher Key. 'Thomas Berry, Buddhism, and the New Cosmology'. *Buddhist-Christian Studies* 18 (1998): pp. 147–54.

Cilliers, Paul. *Complexity and Postmodernism: Understanding Complex Sys-*

tems. 1998. Reprint, London: Routledge, 2005.

Clark, Tim. 'A Whiteheadian Chaosmos: Process Philosophy from a Deleuzean Perspective'. *Process Studies* 3-4 (1999): pp. 179-94.

Cohn, Norman. *Cosmos, Chaos, and the World to Come.* 2nd ed. New Haven: Yale Nota Bene, 2001.

Conklin, Edmund S. 'Superstitious Belief and Practice Among College Students'. *The American Journal of Psychology* 30 (1919): pp. 83-102.

Corcoran, E.'The Edge of Chaos'. *Scientific American* 267.A (1992).

Cornford, Francis M. *Plato's cosmology.* 1935. Reprint, Cambridge: Hackett Publishing Company, 1997.

Curry, Patrick. 'Astrology', pp. 55-57. In *The Encyclopaedia of Historians and Historical Writing 2 Vols.* Edited by Kelly Boyd. London: Fitzroy Dearborn, 1999.

———. 'Memorial to Michel Gauquelin'. *Correlation* 11 (1991).

Davies, P. *The New Physics.* New York: Cambridge University Press, 1989.

de Saint-Périer, R. 'La statuette féminine de Lespugue (Haute-Garonne)'. *Bulletin de la Société préhistorique de France* 21 (1924): pp. 81-84.

Dean, Geoffrey. *Recent Advances in Natal Astrology.* Australia: Fowlers, 1977.

Deleuze, Gilles. *Difference and Repetition.* London: Continuum, 2004.

Descartes, René. *Principles of Philosophy.* 1644. Reprint, New York: Springer, 1984.

Doane, Doris Chase. *Astrology, 30 Years Research.* Los Angeles: The Church of Light, 1956.

Drake, Stillman. *Galileo.* Oxford: Oxford University Press., 1996.

Dunand, Francoise, and Christiane Zivie-Coche. *Gods and Men in Egypt, 3000 BCE to 395 CE.* London: Cornell University Press, 2004.

Ebertin, Reinhold. *The Combination of Stellar Influences.* 1940. Reprint, Tempe, AZ: AFA, 1997.

———. *Directions: Co-Determinants of Fate.* Translated by Linda Kratzsch. Tempe, AZ: AFA, 1976.

Eenwyk, John van. *Archetypes and Strange Attractors: The Chaotic World of Symbols.* Toronto: Inner City Books, 1997.

Eigenauer, John D. 'The Humanities and Chaos Theory: A Response to Steenburg's "Chaos at the marriage of heaven and hell"'. *The Har-*

vard Theological Review 86 (1993): pp. 455–69.

Eliade, Mircea. *The Sacred and the Profane: The Nature of Religion*. 1957.
    Reprint, London: Harcourt, Inc., 1987.

Elwell, Dennis. 'Editorial'. *Correlation* 8 (1970).

Eysenck, Hans. 'The Importance of Methodology in Astrological Re-
    search'. *Correlation* 1 (1981).

Farronato, Cristina. *Eco's Chaosmos: From the Middle Ages to Postmodernity*.
    Toronto: University of Toronto Press, 2003.

———. 'From the Rose to the Flame: Eco's Theory and Fiction Between
    the Middle Ages and Postmodernity', pp. 50–70. In *New essays on
    Umberto Eco*. Edited by Peter E. Bondanella. Cambridge: Cambridge
    University Press, 2009.

Fortune, Dion. *The Mystical Qalabah*. 1935. London: Ernest Benn, 1972.

Forward, Susan, and Craig Buck. *Toxic Parents: Overcoming the Legacy of
    Parental Abuse*. London: Bantam, 1990.

Frankfort, Henri. *Ancient Egyptian Religion*. New York: Columbia Univer-
    sity Press, 1948.

Galilei, Galileo. *Dialogue Concerning the Two Chief World Systems*. 1632.
    Reprint, New York: The Modern Library, 2001.

Gentner, Dedre, et al. 'Analogical Reasoning and Conceptual Change: A
    Case Study of Johannes Kepler'. *The Journal of the Learning Sciences* 6
    (1997): pp. 3–40.

Gettings, Fred. *Dictionary of Astrology*. London: Routledge & Kegan Paul,
    1985.

Gleick, James. *Chaos: Making a New Science*. New York: Viking-Penguin,
    1987.

Goetz, Philip W. 'Bach', pp. 539–44. In *The New Encyclopaedia Britannica*.
    Edited by Philip W. Goetz. Chicago: Encyclopaedia Britannica, Inc..

Greene, Liz. *The Art of Stealing Fire*. London: CPA Press, 1996.

Griffin, David Ray. 'Process philosophy of religion'. *International Journal
    for Philosophy of Religion* 50 (2001): pp. 131–51.

Guirand, F. *Egyptian Mythology*. New York: Tudor, 1965.

Gunzburg, Darrelyn. *Life After Grief: An Astrological Guide to Dealing with
    Loss* (Bournemouth, UK: Wessex Astrologer, 2004).

Hand, Robert. *Planets in Transit*. Rockport, MA: Para Research, 1976.

Hastings, Nancy Anne. *Secondary Progressions: Time to Remember*. York Beach, ME: Samuel Weiser, 1984.

Heraclitus. *Heraclitus Fragments*. New York: Penguin Classics, 2003.

Hesiod. *Works and Days, Theogony and the Shield of Hercules*. New York: Dover Publications Inc., 2006.

Hunger, Hermann. *Astrological Reports to Assyrian Kings*. Helsinki: Helsinki University Press, 1992.

Hunger, Hermann, and David Edwin Pingree. *Astral Sciences in Mesopotamia*. Leiden: Brill, 1999.

Joyce, James. *Finnegans Wake*. Oxford: Oxford University Press, 2012.

Kane, Sean A. *Wisdom of the Mythtellers*. Ontario: Broadview Press, 1998.

Kant, Immanuel. *Kritik der Urteilskraft*. Edited by G. Lehmann. Stuttgart: Reclam, 1971.

Kauffman, Stuart. *At Home in the Universe: The Search for the Laws of Self-Organization and Complexity*. New York: Oxford University Press, 1995.

———. 'Antichaos and adaptation'. *Scientific American* 256 (1991): pp. 78–84.

Keepin, William. 'Astrology and the New Physics'. *The Mountain Astrologer* (Aug/Sept., 1995): pp. 12–21.

Kepler, Johannes, *Astronomia nova*. 1609. Reprint, London: Cambridge University Press., 1992.

King, L. W. *The Seven Tablets of Creation: or, The Babylonian and Assyrian Legends Concerning the Creation of the World and of Mankind*. 1902. Reprint, Montana: Kessinger Publishing, 2004.

Koestler, Arthur. *The Sleepwalkers: A History of Man's Changing Vision of the Universe*. Harmondsworth, UK: Penguin Books Ltd, 1959.

Kuberski, Philip. *Chaosmos: Literature, Science, and Theory*. Albany: State University of New York Press, 1994.

Laëtius, Diogenes. 'Physics', pp. 51–112. In *The Stoics Reader*, ed. by Brad Inwood and Lloyd P. Gerson. Cambridge, MA: Hackett Publishing Company Inc., 2008.

Laplace, Pierre-Simon. *A Philosophical Essay on Probabilities*. 1812. Reprint, New York: Cosmio, 2007.

Larson, Edward J. *Evolution: The Remarkable History of a Scientific Theory*.

New York: Modern Library, 2004.

Lehoux, Daryn. 'Tomorrow's News Today: Astrology, Fate, and the Way Out'. *Representations* 95 (Summer 2006): pp. 105–22.

Leo, Alan. *The Progressed Horoscope*. 1905. Reprint, London: L. N. Fowler, 1929.

Long, A. A., and D. N. Sedley. *The Hellenistic Philosophers*. 1987. Reprint, Cambridge: Cambridge University Press, 2007.

Luce, J. V., *An Introduction to Greek Philosophy*. London: Thames and Hudson Ltd., 1992.

Lull, Ramon. *Treatise on Astronomy Books II–V*. Edited by Kristina Shapar. Translated by Robert Hand. Berkeley Springs, WV: Golden Hind Press, 1994.

Mainzer, Klaus. *Thinking in Complexity: The Complex Dynamics of Matter, Mind and Mankind*. London: Springer-Verlag, 1994.

Mandelbrot, Benoit, *Fractals: Form, Chance, and Dimensions*. San Francisco: W. H. Freeman, 1977.

Mansueto, Anthony. 'Cosmic Teleology and the Crisis of the Sciences'. *Philosophy of Science* (1998) http://www.bu.edu/wcp/Papers/Scie/ScieMans.htm

Masha'allah. *On Reception*. Reston, VA: ARHAT Publications, 1998.

Maturana, Humberto, and Francisco Varela. *The Tree of Knowledge*. Boston, MA: Shambhala, 1987.

McClure, Mary Ann. 'Chaos and Feminism—A Complex Dynamic: Parallels Between Feminist Philosophy of Science and Chaos Theory'. 2004. http://www.pamij.com/feminism.html

Meyer, Michael R. *A Handbook for the Humanistic Astrologer*. New York: Anchor Press, 1974.

Middleton, C., G. Fireman, and R. DiBello. 'Personality Traits as Strange Attractors'. Paper presented at the *Inaugural Meeting for the Society for Chaos Theory in Psychology*, San Fransicso, CA, 1991.

Monmouth, Geoffrey of. *History of the Kings of Britain*. London: J. M. Dent and Co, 1904.

Morin, Jean-Baptiste. *Astrologia Gallica Book Twenty-Two Directions*. Tempe, AZ: AFA, 1994.

Myths, Australian. http://www.dreamscape.com/morgana/miranda.

htm#AUS

Myths, Chinese. http://www.dreamscape.com/morgana/ariel.htm#HAW

NCGR Journal. Edited by Buryl Payne, NCGR. Stamford, CT: The National
    Council for Geocosmic Research, 1986-7.

Newton, Isaac. Principia. Berkeley, CA: University of California Press,
    1966.

Olschki, Leonardo. 'Galileo's Philosophy of Science'. The Philosophical
    Review 52 (1943): pp. 349-65.

Parpola, Simo. Letters from Assyrian Scholars to the Kings Esarhaddon and
    Assurbanipal Part 1. Germany: Butzon and Kevelaer, 1970.

Philipson, Garry. 'Astrology and the Anatomy of Doubt'. The Mountain
    Astrologer 104 (2002): pp. 2-12.

———. Astrology in the Year Zero. London: Flare Publications, 2000.

Plato. Gorgias. In Plato Complete Works. Edited by John M. Cooper. Cam-
    bridge: Hackett Publishing Company, 1997.

———. Republic. In Plato Complete Works. Edited by John M. Cooper. Cam-
    bridge: Hackett Publishing Company, 1997.

———. Timaeus. In Plato Complete Works. Edited by John M. Cooper. Cam-
    bridge: Hackett Publishing Company, 1997.

Plotinus. The Enneads. Translated by Steven MacKenna. New York: Lar-
    son Publications, 1992.

Poincaré, Henri Jules. Science and Method. London: T- Nelson, 1914.

Polley, Douglas. 'Turbulence in Organizations: New Metaphors for Orga-
    nizational Research'. Organization Science 8 (1997): pp. 445-57.

Poole, Robert. 'Is It Chaos, or Is It Just Noise?'. Science 243 (6 January
    1989): pp. 25-28.

Prigogine, Ilya, and Isabelle Stengers. Order Out of Chaos: Man's New Dia-
    logue with Nature. New York: Bantam Books, 1984.

Ptolemy, Claudius. The Tetrabiblos. Mokelumne Hill, CA: Health Research,
    1969.

Reiner, Erica. 'Babylonian Celestial Divination', 21-37. In Ancient Astrono-
    my and Celestial Divination. Edited by N. M. Swerdlow. London: MIT
    Press, 1999.

Reston, James Jr. Galileo: A Life. New York: Harper Collins, 1994.

Rochberg-Halton, Francesca. 'Elements of the Babylonian Contribution to Hellenistic Astrology'. *Journal of the American Oriental Society* 108 (1988): pp. 51–62.

Rudhyar, Dane. *The Lunation Cycle*. Santa Fe, NM: Aurora Press, 1967.

Ruf, Frederick J. *The Creation of Chaos: William James and the Stylistic Making of a Disorderly World*. Albany: State University of New York Press, 1991.

Schlegel, Friedrich von, *The Philosophy of Life and Philosophy of Language in a Course of Lectures*. London: Henry G. Bohn, 1847.

Shallis, Michael. 'Problems of Astrological Research'. *Correlation* 1 (1981).

Shulman, Helene. *Living on the Edge of Chaos: Complex Systems in Culture and Psyche*. Zurich: Daimon, 1997.

Smit, Rudolf. 'Editorial'. *Correlation* 14 (1995).

Solomon, Robert. 'On Fate and Fatalism'. *Philosophy East and West* 53 (2003): pp. 435–54.

Stacey, Ralph D. *Complex Responsive Processes in Organizations: Learning and Knowledge Creation*. London: Routledge, 2001.

Stewart, Ian. *What Shape is a Snowflake?* New York: The Ivy Press Limited, 2001.

Stone, Merlin. *Ancient Mirrors of Womanhood: Our Goddess and Heroine Heritage. Vol. 1*. New York: New Sibylline Books, 1979.

Straus, Murray A., and Richard Gelles. *Physical Violence in American Families: Risk Factors and Adaptations to Violence in 8,145 Families*. New Brunswick: Transaction Publishers, 1990.

Strogatz, Steven H. *Nonlinear Dynamics and Chaos: With Applications in Physics, Biology, Chemistry, and Engineering*. Reading, MA: Addison-Wesley, 1994.

Stuart, Dorothy Margaret. *Book of Chivalry and Romance*. London: G. G. Harrap, 1933.

Taub, Liba Chaia. *Ptolemy's Universe: The Natural Philosophical and Ethical Foundations of Ptolemy's Astronomy*. Chicago: Open Court, 1993.

Tester, James. *A History of Western Astrology*. Woodbridge, UK: The Boydell Press, 1987.

Thiétart, R. A., and B. Forgues. 'Chaos Theory and Organization'. *Organization Science* 6 (1995): pp. 19–31.

Waldrop, Mitchell. *Complexity: The Emerging Science at the Edge of Order and Chaos*. New York: Touchstone, 1992.

Walker, C. B. F. 'A Sketch of the Development of Mesopotamian Astrology and Horoscopes', pp. 7–14. In *History and Astrology*. Edited by Annabella Kitson. London: Uwin Paperbacks, 1989.

Ward, Margaret. 'Butterflies and Bifurcations: Can Chaos Theory Contribute to our Understanding of Family Systems?'. *Journal of Marriage and the Family* 57 (August 1995): pp. 629–38.

Williams, Bernard. 'Plato—The Invention of Philosophy', pp. 47–92. In *The Great Philosophers*. Edited by Ray Monk and Frederic Raphael. London: Phoenix, 2001.

Willis, Roy, and Patrick Curry. *Astrology, Science and Culture: Pulling Down the Moon*. New York: Berg, 2004.

Woodruff, John, Neil Bone, and Storm Dunlop. *Philips' Astronomy Dictionary: An Illustrated A-Z Guide to the Universe*. London: George Philip Ltd, 1995.

Index

SOPHIA CENTRE PRESS